Surveys and Tutorials in the Applied Mathematical Sciences

Volume 7

Featuring short books of approximately 80-200pp, Surveys and Tutorials in the Applied Mathematical Sciences (STAMS) focuses on emerging topics, with an emphasis on emerging mathematical and computational techniques that are proving relevant in the physical, biological sciences and social sciences. STAMS also includes expository texts describing innovative applications or recent developments in more classical mathematical and computational methods.

This series is aimed at graduate students and researchers across the mathematical sciences. Contributions are intended to be accessible to a broad audience, featuring clear exposition, a lively tutorial style, and pointers to the literature for further study. In some cases a volume can serve as a preliminary version of a fuller and more comprehensive book.

More information about this series at http://www.springer.com/series/7219

Haiyan Wang · Feng Wang · Kuai Xu

Modeling Information Diffusion in Online Social Networks with Partial Differential Equations

 Springer

Haiyan Wang
School of Mathematical & Natural Sciences
Arizona State University
Phoenix, AZ, USA

Feng Wang
School of Mathematical & Natural Sciences
Arizona State University
Phoenix, AZ, USA

Kuai Xu
School of Mathematical & Natural Sciences
Arizona State University
Phoenix, AZ, USA

ISSN 2199-4765   ISSN 2199-4773   (electronic)
Surveys and Tutorials in the Applied Mathematical Sciences
ISBN 978-3-030-38850-8   ISBN 978-3-030-38852-2   (eBook)
https://doi.org/10.1007/978-3-030-38852-2

Mathematics Subject Classification (2010): 05C50, 05C60, 34A34, 35-02, 35C07, 35R35, 35B36, 35K57, 35B32, 35Q94, 35R02, 62H30, 65F15, 68-02, 68P01, 68R10, 68U35, 91C20, 92D25, 91D30, 94A15, 94C15

This Springer imprint is published by the registered company Springer Nature Switzerland AG.
The registered company address is: Gewerbestrasse 11, 6330 Cham, Switzerland

To our families.

Preface

For many researchers, the rise of tremendously influential social media such as Twitter and Facebook presents both challenges and opportunities. For example, a better understanding of the information diffusion process over online social networks can effectively predict and coordinate online social activities. However, even though the increasing availability of unprecedented quantities of data has accelerated research on information diffusion in online social networks, because of the complexity of social interactions and rapid changes in social media, the mechanism of information diffusion in online social networks remains elusive. In the literature various mathematical models have been proposed to study information diffusion on online social networks. However, these mathematical models, particularly, deterministic dynamical models, are largely based on ordinary differential equations (ODEs) that deal with collective social processes over time.

The goal of this book is to introduce a new dynamic modeling approach to the use of partial differential equations (PDEs) for describing temporal-spatial patterns in information diffusion over social media. The PDE-based models are reaction-diffusion equations built on intuitive social distances between communities (clusters) of online users. In particular, leveraging clustering analysis of spatial big data will dramatically expand applications of the PDE models to many health and social problems including influenza prediction. The PDE approach advocates a paradigm shift for modeling information diffusion in online social networks and lays the theoretical groundwork for many spatio-temporal modeling problems in the big data era.

This book stems from lecture notes for a course, Social Media and Mathematics, given by the first author, at Arizona State University and other institutions, and the authors' numerous published and unpublished works on modeling information diffusion over online social networks with PDEs.

Background preparations and necessary references for social graphs and information diffusion are also included to ensure the book is accessible to mathematicians and computer scientists as well as general researchers in social media.

Phoenix, AZ, USA   Haiyan Wang
Phoenix, AZ, USA   Feng Wang
Phoenix, AZ, USA   Kuai Xu
October 29, 2019

Acknowledgments

We would like to acknowledge the contributions to the papers from our collaborators: Guowei Dai, Xiaohua Jia, K. Hazel Kwon, Chengxia Lei, Zhigui Lin, Ruyun Ma, Chuan Peng, Jingli Ren, Yufang Wang, Jianhong Wu, Hongyong Zhao, Shuhua Zhang, Dandan Zhu, Linhe Zhu. We would thank several colleagues including Guoping Jiang, Zhen Jin, Maoxing Liu, Yuan Lou, Yurong Song, and a number of students for stimulating discussions. Adrian Avram, Jaime Chon, Ross Raymond, Paul Wagenseller, Daniel Langley, Shaun Fuller, and a number of students helped Twitter data collection and Matlab simulations. Dingyong Bai and Xiaoling Han helped organize the references. We would like to thank Kristina Lerman for making the Digg 2009 data set available to the research project.

We want to thank the reviewers for their inputs as well as the editorial staff of Springer who were involved in the oversight of the book. The Springer editors encouraged and supported the project from the beginning. We are grateful for their encouragement and support.

Finally, we would like to acknowledge that this project has been supported by two research grants from the U.S. National Science Foundation (CNS-1218212 and DMS-1737861).

Contents

1   **Introduction**   1

2   **Ordinary Differential Equation Models on Social**
     Networks   3
    2.1   Introduction   3
    2.2   Diffusion of Innovations   4
        2.2.1   Characteristics of Innovation Diffusion   4
        2.2.2   Innovation Diffusion Models   8
    2.3   Epidemical Model   10
        2.3.1 SIR Model   11
        2.3.2 SIS Model   12
        2.3.3 SIR Model with Standard Incidence   12

3   **Spatio-Temporal Patterns of Information Diffusion**   15
    3.1   Introduction   15
    3.2   Digg Data Studies   15
    3.3   Friendship Hops as Distance   17
        3.3.1   Friendship Hops   17
        3.3.2   Logistic Influence Curve   19
    3.4   Shared Interests as Distance   19
        3.4.1   Shared Interests   19
        3.4.2   Logistic Influence Curve   20
        3.4.3   Studies of Shared Interests   20

4   **Clustering of Online Social Network Graphs**   27
    4.1   Introduction   27
    4.2   Graph Models of Online Social Networks   28
    4.3   Spectral Graph Bipartitioning   30
    4.4   Clustering Based on Higher-Order Organization   32
    4.5   Spectral Partitioning with Bipartite Graph   37
    4.6   Discussions   40

5   **Partial Differential Equation Models**   43
    5.1   Introduction   43
    5.2   Embedding Network Graphs to Euclidean Spaces   44
    5.3   External and Internal Influences   45
    5.4   PDE Model Formulation   47
    5.5   Diffusive Logistic Model   50
          5.5.1   Initial Density Function Construction   52
          5.5.2   Accuracy of Diffusive Logistic Model   52
    5.6   Linear Diffusive Model   53
          5.6.1   Accuracy of Linear Model   54
    5.7   Logistic Model with Biased Diffusion   56
          5.7.1   Accuracy of Logistic Model with Biased Diffusion ...   58

6   **Modeling Complex Interactions**   59
    6.1   Introduction   59
    6.2   Information Diffusion Initiated from Multiple Sources   60
          6.2.1   Distance Metric   60
          6.2.2   Experiment Results   61
    6.3   Cooperating Diffusion Process   63
          6.3.1   Cooperative Systems   63
          6.3.2   Modeling the Interaction of Mobile Phones and
                  Mobile Applications   64
    6.4   Competing Diffusion Process   65
          6.4.1   Competition Systems   65
          6.4.2   Modeling the Ecosystem of Smartphone Operating
                  Systems   66
    6.5   Spatial Epidemiological Models   67
    6.6   Multiple Communication Channels   68

7   **Mathematical Analysis**   69
    7.1   Introduction   69
    7.2   Free Boundary Problems in Online Social Networks   70
          7.2.1   Free Boundary Problems   70
          7.2.2   Free Boundary Problems with Multiple Information ..   71
          7.2.3   Theoretical Results on Free Boundary Problems   73
          7.2.4   Discussion   81
    7.3   Stability and Bifurcation   82
          7.3.1   A More General Boundary Condition   82
          7.3.2   Upper and Lower Solutions   83
          7.3.3   Eigenvalue Problems   85
          7.3.4   Information Spreading and Vanishing   87
          7.3.5   Discussion   90
    7.4   Hopf Bifurcation of an Epidemic-Like Rumor Model   91
          7.4.1   Mathematical Analysis and Simulation   92
          7.4.2   Discussion   98

        7.5   Traveling Wave Solutions and Spreading Speeds   98
               7.5.1   Long-Term Propagation of Information on Social
                        Networks   98
               7.5.2   Mathematical Formulation and Assumptions   99
               7.5.3   Spreading Speed and Traveling Wave Solutions   103
               7.5.4   Propagation Speeds of Cooperative Information   107
               7.5.5   Propagation Speeds for Competing Information   108
               7.5.6   Propagation Speeds for Spatial Epidemiological
                        Models   109
               7.5.7   Discussion   112

8   Applications   113
        8.1   Introduction   113
        8.2   Analysis of Twitter Information Diffusion During the
               Egyptian Revolution   113
               8.2.1   The 2011 Egyptian Revolution and Social Media   113
               8.2.2   Data Collection   114
               8.2.3   PDE Modeling of Global Information Diffusion   117
               8.2.4   Model Validation   121
               8.2.5   Discussion   122
        8.3   A PDE-Based Influenza Surveillance System   123
               8.3.1   Twitter Data   124
               8.3.2   PDE Modeling Based on Higher-Order Graph
                        Clustering   126
               8.3.3   PDE Modeling Based on Bipartite Graph Clustering .   129
               8.3.4   Discussion   131

References   133

Index   143

Chapter 1
Introduction

Abstract Online social networks (OSNs) such as Twitter and Facebook, emerging as the "model organism" of Big Data, have gained tremendous popularity for the platforms they provided for information exchange. Much of prior work on information diffusion over online social networks has been based on empirical and statistical approaches. The majority of dynamical models arising from information diffusion over online social networks are ordinary differential equations (ODEs). Recently, the authors proposed to use partial differential equations (PDEs) to model information diffusion in online social networks and introduced a new transdisciplinary architecture for modeling information diffusion. These studies demonstrate fascinating connections between advanced mathematics and online social networks.

Online social networks (OSNs) such as Twitter and Facebook, emerging as the "model organism" of Big Data, have gained tremendous popularity for the platforms they provided for information exchange. Much of prior work on information diffusion over online social networks has been based on empirical and statistical approaches. The majority of dynamical models arising from information diffusion over online social networks are ordinary differential equations (ODEs). Recently, the authors proposed to use partial differential equations (PDEs) to model information diffusion in online social networks and introduced a new transdisciplinary architecture for modeling information diffusion. These studies demonstrate fascinating connections between advanced mathematics and online social networks.

A significant body of research about online social networks has focused on analysis of such networks with empirical approaches that use data mining and statistical modeling schemes [9, 18, 19, 25, 29, 34, 35, 41, 43, 47, 50, 55, 60–62, 68, 69, 72, 73, 85, 88, 89, 105, 111, 120, 121, 138–141, 144]. Mathematical models have played a significant role in understanding and predicting

© Springer Nature Switzerland AG 2020   1
H. Wang et al., *Modeling Information Diffusion in Online Social Networks with Partial Differential Equations*, Surveys and Tutorials in the Applied Mathematical Sciences 7, https://doi.org/10.1007/978-3-030-38852-2_1

information diffusion in online social networks over time. In particular, epidemiological models have influenced the research on information diffusion [8, 51, 80, 89, 90, 124, 145, 146]. However, the deterministic models proposed for online social networks in the literature are largely based on ordinary differential equations (ODEs) that deal with collective social processes over time.

In a paper [128], the authors proposed to use partial differential equations (PDEs) built on intuitive cyber-distance among online users to study both temporal and spatial patterns of information diffusion process in social media. One of the simple, yet fundamental questions that the models address is this: for a piece of given information m initiated from a particular user called *source* s, what is the density of influenced users at network distance x from the source at *any time* t. We validate the models with real datasets collected from two popular social media sites, Twitter and Digg. The experiment results show that the models can achieve over 90% accuracy and effectively predict the density of influenced users.

This paper [128] is the first attempt to propose a PDE-based model for characterizing and predicting the temporal and spatial patterns of information diffusion over online social networks, which is also indicated in a survey by Zhang et al. [146]. In Guille et al.'s survey [38] on information diffusion over online social networks, the PDE model in [128] is reported as one of the three non-graph predictive models: epidemiological models, linear influence model (LIM), and PDE approach. The LIM approach developed in [140] focuses on predicting the temporal dynamics of information diffusion through solving non-negative least squares problems. The PDE-based models are dynamic systems that take into account the influence of the underlying network structure as well as information contents for predicting information diffusion over both temporal and spatial dimensions. We shall propose a number of spatio-temporal epidemiological models to describe information diffusion in online social networks in this book.

The book lies at the interface of mathematics, social media analysis, network science, and data science. A key challenge of this interdisciplinary research is to integrate big data from online social networks into the framework of partial differential equations. We use clustering analysis from data mining to aggregate big data for the validation of the PDE models. There are many clustering methods such as k-means clustering. We focus on spectral clustering analysis in this book as our data are graph-structured. The book integrates the research efforts carried out collaboratively by mathematicians specialized in partial differential equations, computer scientists focused on network theory and data mining, and researchers in social media. The extension of partial differential equations into online social networks presents new opportunities and challenges for mathematicians as well as computer scientists and researchers in social media.

Chapter 2
Ordinary Differential Equation Models on Social Networks

Abstract In this chapter we consider a number of ordinary differential equation models for diffusion of innovation and epidemiological models. We discuss the classical theory on diffusion of innovation, emphasizing online social networks and analyzing several ordinary differential equation models for innovation diffusion. We also present a number of basic compartment epidemiological models and their applications in online social networks, and finally we discuss SIR models and their extensions when the total population is not constant.

2.1 Introduction

Information diffusion over online social networks has become a fast growing research domain encompassing techniques from a plethora of sciences, among them mathematics, computer science, communications and marketing, etc. In addition, the method for studying the spread of infectious diseases among a population has been applied to understanding information spreading patterns. Information diffusion has become a subject with disparate views of what is an information diffusion process. We focus here on a particular case in which information diffuses in online social networks, and we define information diffusion as the process by which a piece of information (knowledge) spreads and reaches individuals through interactions in a network.

The theory of information diffusion can be traced back to the research on the diffusion of innovation over a population, advanced by a pioneering mass communication scholar, E. M. Rogers. Information diffusion in online social networks becomes a premier example to revisit innovation diffusion. In this chapter, we will focus on ODE models for diffusion of innovation

© Springer Nature Switzerland AG 2020

H. Wang et al., *Modeling Information Diffusion in Online Social Networks with Partial Differential Equations*, Surveys and Tutorials in the Applied Mathematical Sciences 7, https://doi.org/10.1007/978-3-030-38852-2_2

and epidemics. Both diffusion of innovations and epidemic models provide a global view of how an innovation (e.g., news, a product) or a type of disease spreads through a population even when interactions among individuals are unavailable. We will extend innovation diffusion and epidemic models to spatial models where both local and global information between clusters in a network are available.

2.2 Diffusion of Innovations

Diffusion of innovations is a theory that seeks to explain how new ideas and technologies spread through cultures. E. M. Rogers was a professor of communication studies who popularized the theory with his book "Diffusion of Innovations" [104]. The origins of the diffusion of innovations theory span various disciplines including anthropology, early sociology, rural sociology, education, industrial sociology, and medical sociology. Rogers [104] theorized that information diffusion is the social process through which an innovation is communicated through certain channels over time among the participants in a social system. He identified four key elements that influence diffusion of a new idea: the innovation itself, communication channels, time, and a social system. The theory of diffusion of innovation has been applied to numerous contexts, including marketing, communications, health promotion, organizational studies, and complexity studies.

The rise of social media has provided a new platform to study diffusion of innovation. Rogers [104] defined an innovation as an idea, practice, or object if it is perceived as novel by an individual or other unit of adoption. A piece of news being reposted many times in online social networks is a typical example of innovations diffusing across online social networks. The theory of diffusion of innovations can be applied to online social networks to answer why and how information spreads fast and reaches a broad audience. It also can be used to reveal some key characteristics of information diffusion such as its motivations. With appropriate models, it can be used to further predict the rate at which ideas spread. In this section, we review some key characteristics of innovation diffusion that are important for the PDE modeling of information diffusion in online social networks. Finally, we present mathematical models that can be used to describe the process of innovation diffusion.

2.2.1 Characteristics of Innovation Diffusion

Rogers [104] explored many characteristics of innovations, individual adopters, and organizations. The nature of networks and the roles opinion leaders play determine the likelihood of innovation adoption. We here identify several characteristics that are related to mathematical modeling of information diffusion in online social networks. Our goal is to use mathematical models to identify key factors that influence the spread of information, and then explain how and why users adopt certain information.

2.2.1.1 Logistic Adoption Curve

In [104] Rogers argued that the diffusion process, depending on a combination of the four key elements, has a point at which an innovation reaches critical mass. He then studied the various stages and introduced an adopter category as a classification of individuals within a social system on the basis of innovativeness. Based on the order in which they adopt the innovations, it has been found that the five different types of adopters are: (1) Innovators (top 2.5%), (2) Early Adopters (13.5%), (3) Early Majority (34%), (4) Late Majority (34%), and (5) Laggards (16%). Numerous studies have shown that different types of adopters behave in significantly different ways in various cultures and fields. Figure 2.1 shows the distribution of the five categories of adopters as well as the cumulative adoption S-shaped curve.

We also plot the cumulative adoption time path or temporal pattern of a diffusion process over time in this figure. The resulting distribution over time can generally be described as taking the form of an S-shaped (sigmoid) curve. The S-shaped (sigmoid) curve indicates that the adoption rate is relatively slow when innovators or early adopters begin to adopt the product or information. As soon as the majority of early adopters are onboard, the adoption curve becomes linear, and the rate is close to constant until all late majority members begin to adopt the product. After the late majority adopts the product, the adoption rate gradually becomes slow once again when laggards start adopting. In the end, the curve slowly approaches 100%.

Such a curve is also called a *logistic curve*, a term that has been used in different fields such as biological growth. From mathematical point of view, there is an inflection point where the curve changes from concave-down to concave-up. Before the infection point, only a small numbers of the social system slowly adopt the innovation in each time period. After the inflection point, the number of adoptions per time period accelerates as the diffusion process begins to expand more in a community. Ultimately the curve reaches an upper asymptote, which often is the capacity of the network.

A wide spectrum of innovation diffusion processes have confirmed that the adoption rates for different diffusion processes follow S-shaped curves. The diffusion pattern of most innovations follows a general S-shaped curve. However, the exact form of each curve, including the slope and the asymptote, may exhibit differences. For example, depending on specific information or production, the slope may be very steep initially, indicating rapid diffusion, or it may be gradual, indicating relatively slow diffusion. Figure 2.3 shows two different S-shaped curves for internal-influence diffusion and mixed-influence diffusion with two different increasing rates.

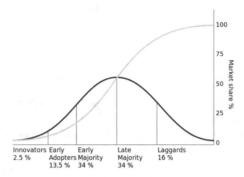

Fig. 2.1 The diffusion of innovations according to Rogers. As successive groups of consumers adopting the new technology (shown in blue), its market share (yellow) will eventually reach the saturation level. In mathematics, the yellow curve is known as the logistic function. The curve is broken into sections of adopters

Diffusion research centers on identifying the conditions to facilitate the increase or decrease of the likelihood that a new idea, product, or practice will be adopted by members of a given culture. More importantly, it is of great interest to predict how fast innovations spread and how media as well as interpersonal contacts influence opinion and judgment. Mathematical models based on the logistic adoption curve and five categories of adopters may provide guidance for taking appropriate actions to promote or control information spreading in online social networks.

2.2.1.2 Opinion Leaders and Clustering

It is evident that individuals in a population do not play an equal amount of influence over others. Opinion leaders, who have more social experience and higher social status, are more influential in convincing other people to change their attitudes toward news or innovations. They, in general, are able to have access to the mass media and their influence in spreading either positive or negative information is larger than that of other people. The two-step flow theory was originated in [53, 54] for studying mass communications. The two-step flow theory reveals that information diffusion propagates in two distinct stages. First, opinion leaders are willing to accept the information. Second, opinion leaders pass on their own interpretations or opinions to others. As a result, the influence of opinion leaders is larger than others. The process is illustrated in Fig. 2.2. The two-step flow theory also has been extended to the multi-step flow theory for mass communication and innovation diffusion.

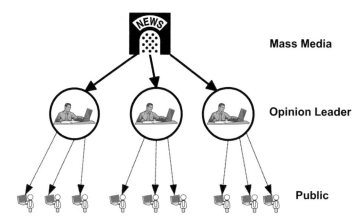

Fig. 2.2 Two-step flow

One of the successful applications of the two-step flow theory is to design marketing strategies to maximize the efficiency of broadcast advertising on the diffusion of new products and services. Opinion leaders are early adopters and then pass positive information on new products to a large number of other people through various channels such as online reviews. It was found [99] that opinion leadership tended to be organized into a hierarchical structure within a society as shown in Fig. 2.2. Within each level in the hierarchy, the opinion leaders have the most influence over other members. The bottom levels have generally larger numbers of people. Opinion leaders often are targeted by mass advertising. The two-step flow theory indicates that interpersonal communications are also effective as well. In addition, advertising should best target on those who may adopt. However, it is often difficult to find those who may adopt a product. Today, big data analytics could provide information for identifying groups of people who may be interested in a product.

In recent years, social media such as Facebook, Twitter, and hundreds of blog sites provide new platforms for people to share information in online social networks. The increasing impact of social media provides a new perspective and creates huge amount of data to study the role of opinion leaders in social media. In social media key opinion leaders include influential politicians, famous researchers, journalists, bloggers, and others who are actives on more than one social media platforms. However, it is found that the person with the most followers is not necessary the opinion leaders in social media. It is often difficult to discern the opinion leaders who have the influences on others. Various techniques have been used to identify opinion leaders in social media.

Social communities/clusters are ubiquitous, occurring in any social networks based on common backgrounds such as families, politics and business, colleges, cities, states, nations, continents. Many communities and clusters have a tendency to become spatially concentrated at the regional level, in part because of the significant role of opinion leaders. The spread of information in social media also exhibits clustering patterns due to other reasons such as common interests. Clusters provide stable characteristics and show a clear global view of interactions while individuals often seek their own interests. There are a number of works to study and determine structured location information in Twitter [26]. Wayant et al. [133] studied the global distribution of geolocated tweets over a period of 2 days (November 16th and 17th, 2011) during the event of the Occupy Wall Street movement in New York City, NY, and reveal that the tweet distribution has the highest concentration in the USA and Western Europe, but also spreads as far as Brazil, the Arab peninsula, and Australia.

The diversity of online users in social media makes the structures of communities in social media more complex. With a dynamic and evolving environment in social media, it is often challenging to follow user interactions and identify their communities. Communities and clusters are essential for studying information diffusion in social media. We will use mathematical models based on partial differential equations to characterize information diffusion patterns in social media.

2.2.2 Innovation Diffusion Models

Mathematical models are capable of representing the level or spread of an innovation among a given set of prospective adopters in a social system with respective to time. The advantage of differential equation models is to give the successive increase in the number of adopters or adopting units over time. Thus, differential equation models permit prediction of the continued development of the diffusion process over time within a fixed population and further facilitate a theoretical insight of the dynamics of the diffusion process. Mahajan [78] presents a number of ordinary different equation models to study the diffusion of innovations based on the early research on diffusion processes. These models focus on describing observed diffusion patterns with an assumption that prespecified trends or distribution functions are given. In particular, cumulative normal and logistic adoption curve have been used to model diffusion processes. These models indicate the S-shaped growth curves which is in agreement with the theory of innovation diffusion. Extensive research has been attempted to expand these "diffusion models" to analyze and model the spread of an innovation in various fields. The basic innovation diffusion model can be viewed as an extension of the logistic models in population dynamics [82].

The model takes the following form:

$$\frac{dN(t)}{dt} = g(N,t)[\bar{N} - N(t)] \tag{2.2.1}$$

with the initial condition $N(0) = N_0$, here $N(t)$ is the cumulative number of adopters at time t in a network. $\bar{N}$ denotes the total number of potential adopters. $g(N, t)$ denotes the coefficient of diffusion, which describes the innovativeness of the information or innovation. In general, $g(N, t)$ is a function of the current number of adopter N and time t. This ordinary differential equation quantifies the theory of innovation diffusion, in particular, the S-shaped adoption curve. It states that the rate of diffusion of an innovation at any time t is directly proportional to the gap or difference between the total number of potential adopters and the number of previous adopters at that time. It reveals that the rate at which the number of adopters changes with time depends on the innovativeness of the innovation or information. The diffusion model presented in Eq. (2.2.1) is a deterministic ordinary differential equation. Fortunately, the equation has a closed-form solution $N(t)$, which is the cumulative number of prior adopters; this number eventually approaches the total number of possible adopters in the social system. The peak of the solution is also called the carrying capacity. The coefficient of diffusion, $g(N, t)$, a direct function of N and t, are dependent on many characteristics of the diffusion process such as the nature of the innovation, communication channels, and social system attributes. Mathematically, $g(N, t)$ can be expanded in terms of N

$$g(N, t) = \alpha + \beta N(t) + \gamma N^2(t) + \ldots . \tag{2.2.2}$$

For $g(N, t) = \alpha$, the coefficient of diffusion is independent of the current number of adopters, and will be referred to as the *external-influence diffusion model*. Solving (2.2.2), we can get

$$N(t) = \bar{N} - (\bar{N} - N_0)e^{-\alpha t}.$$

In general, this is not a realistic model as it does not take into account influence within the network. If $g(N, t) = \beta N(t)$, the coefficient of diffusion is linearly proportional to the current number of adopters. The fundamental diffusion model is called the *internal-influence diffusion model*. This is a more realistic model. For the internal-influence diffusion model, by solving (2.2.2), we have

$$N(t) = \frac{\bar{N}}{1 + \frac{\bar{N} - N_0}{N_0} e^{-\beta \bar{N} t}}.$$

Finally, if $g(N, t) = \alpha + \beta N(t)$, the fundamental diffusion model can be expressed as the *mixed-influence diffusion model* (combining external and internal influence). Here α is a parameter to represent external influence and β for internal influence. For the external-influence diffusion model, by solving (2.2.2), we get

$$N(t) = \frac{\bar{N} - \frac{\alpha(\bar{N} - N_0)}{(\alpha + \beta N_0)} e^{-(\alpha + \beta \bar{N})t}}{1 + \frac{\beta(\bar{N} - N_0)}{\alpha + \beta N_0} e^{-(\alpha + \beta \bar{N})t}}.$$

Figure 2.3a, b are the graphs of $N(t)$ with internal-influence diffusion and mixed-influence diffusion. The two curves are similar to the market share curve (yellow) in Fig. 2.1. Note that the curve in Fig. 2.3b (mixed-influence diffusion) increases faster than that in Fig. 2.3a (internal-influence diffusion) due to the fact that external influence accelerates diffusion.

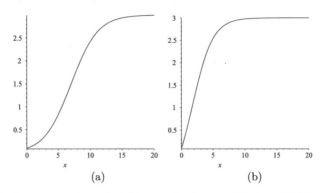

(a)   (b)

Fig. 2.3 Diffusion growth curve. (**a**) Internal-influence diffusion. (**b**) Mixed-influence miffusion

2.3 Epidemical Model

Epidemics of infectious disease have frequently affected our lives. The spread of infectious diseases among a population has been studied for centuries in efforts to understand spreading patterns, such as trends and ratios of people getting infected and therefore to implement effective prevention strategies. Compartmental epidemic models describing the transmission of communicable diseases have been extensively studied, and have many applications to emerging fields such as socio-biological systems and social media. Epidemic models have inspired many unsolved mathematical questions and re-energized novel mathematical research [14]. Conclusions and methods of mathematical modeling of infectious diseases have been accepted as a matter of general interest. More information can be found in [15, 42, 48, 80, 82].

In the present century, deterministic epidemic models have been increasingly adopted to describe and predict computer virus infections and information propagation. There are two categories of epidemics models: those using ordinary differential equations and those using partial differential equations. For ordinary differential equation models, epidemic models assume an implicit network and unknown connections among individuals. Similar to diffusion of innovations models, ordinary differential models are more suitable for studying global patterns, such as trends and ratios of people getting infected, and not interested in who infects whom. Partial differential equation models have spatial factors involved. Here we review two basic epidemic models and one SIR model with standard incidence. In particular, we will discuss some extensions of the models in information diffusion over online social networks.

2.3.1 SIR Model

The spread of infectious diseases often is a complex phenomenon with many interacting factors, and consists of a pathogen being spread, a population of hosts such as humans, and a spreading mechanism. The same components are required for news or rumors to spread. Compartmental epidemic models abstract the population into compartments under certain assumptions. These compartments represent their health status with respect to the pathogen in the system. Let N define the size of this population. Any member of the population can be in either one of three states: (1) Susceptible ($S(t)$), the number of the susceptibles at time t as $S(t)$. (2) Infected ($I(t)$), the number of infected individuals at time t. (3) Recovered (or Removed) ($R(t)$), the size of this set at time t. When an individual is in the susceptible state, he or she can potentially get infected by the disease. An infected individual has the chance of infecting susceptible parties. Individuals can only get infected by infected people in the population. The recovered are individuals who have either recovered from the disease and hence have complete or partial immunity against the infection or were killed by the infection. The simple Kermack–McKendrick model [57] is the starting point for many epidemic models. In a closed population, which consists of susceptible individuals ($S(t)$), infected individuals ($I(t)$), and removed individuals ($R(t)$), the simple deterministic susceptible-infected-removed (SIR) is

$$S' = -\beta SI \tag{2.3.1}$$
$$I' = \beta SI - \gamma I \tag{2.3.2}$$
$$R' = \gamma I, \tag{2.3.3}$$

where β is the transmission coefficient, γ is the recovery rate. It is clear that total population $N = S + I + R$ is constant. The underlying assumptions are that the system has a high level of uncertainty, and individuals usually do not decide whether to get infected or not. In addition, it is assumed that no contact network information is available and the process by which hosts get infected is unknown. In general, a complete understanding of the epidemic process requires substantial knowledge of the biological process within each host. Various extensions of SIR models have been extensively studied on general networks (see, e.g., [89, 145]) and online social networks as well [144].

For online social networks, S can stand for "susceptible," I for "infected" (i.e., adopted the information), and R for "recovered" (i.e., refractory). The model can predict the various properties of the information spread, for example, the prevalence (total number of infected from the epidemic) and the duration of the epidemic. In addition, it also helps understand how different environments may affect the outcome of information diffusion.

2.3.2 SIS Model

The SIS model considers a fixed population with only two compartments susceptible $S(t)$ and infected $I(t)$, thus the flow of this model may be considered as follows:

$$S \to I \to S.$$

The SIS can be easily derived from the SIR model by simply considering that the individuals recovered with no immunity to the disease, that is, individuals are immediately susceptible once they have recovered. Thus removing the equation representing the recovered population from the SIR model and adding those removed from the infected population into the susceptible population, we can get the following differential equations:

$$S' = -\beta SI + \gamma I, \tag{2.3.4}$$
$$I' = \beta SI - \gamma I. \tag{2.3.5}$$

Leskovec et al. [69] used a simple and intuitive SIS model with a single parameter, β, to model cascading behavior in large blog graphs. Each blog is in one of two states: I or S. If a blog is in state I, this means that the blogger just posted a post, and the blog now has a chance to spread its influence. The cascades generated from the model match several power-law properties of real cascades. However, the assumption in the paper that all nodes have the same probability β to adopt the information is not realistic since in real-world social networks, influence is not evenly distributed between all nodes. Thus it is necessary to develop more complex modeling that takes into account this characteristic. The partial differential models in the next few sections will be able to address the issue with different infection rates for different clusters/communities.

2.3.3 SIR Model with Standard Incidence

It is reasonable to assume that the population is closed and fixed when modeling epidemics where the disease spreads quickly in the population and dies out within a short time. In reality, infections can come from outside the population where the disease is being spread (e.g., by genetic mutation, contact with an animal, etc.). Further if the human or animal population growth or decrease is significant or the disease causes enough deaths to influence the population size, then it is not reasonable to assume that the population size is constant (Mena-Lorca and Hethcote [80]). To account for variable population sizes, many researchers have proposed epidemic models with the transmission coefficient taking the following form:

$$\tilde{\beta}(N) = \frac{C(N)}{N},$$ (2.3.6)

where $N = S + I + R$ is the total population size and $C(N)$ is the adequate contact rate. Then mass-action incidence corresponds to the choice $C(N) = \beta N$ and standard incidence corresponds to $C(N) = \beta$, where β is a positive constant. In general, $C(N)$ is a non-decreasing function with respect to N. For example,

$$C(N) = \frac{aN}{1 + bN + \sqrt{1 + 2bN}}$$

(Heesterbeek and Metz [42]),

$$C(N) = \lambda N^{\alpha}, \alpha = 0.05$$

(Mena-Lorca and Hethcote [80]). Other types of $C(N)$ can be found in [14] and references therein.

In Wang et al. [127], a general diffusive Kermack–McKendrick SIR model was studied with the assumption that some of the infective individuals will be removed from the population by disease-induced death or quarantine, but the recovered individuals will return to the community. The total population is $N = S + I + R$ and is not fixed. In this case, we have the following Kermack–McKendrick SIR model with standard incidence:

$$\frac{dS}{dt} = -\frac{\beta S I}{N},$$ (2.3.7)

$$\frac{dI}{dt} = \frac{\beta S I}{N} - \gamma I - \delta I,$$ (2.3.8)

$$\frac{dR}{dt} = \gamma I.$$ (2.3.9)

SIR models with standard incidence may be more suitable for studying information diffusion in online social networks because of rapid changes of online user dynamics. We will discuss propagation speeds for spatial epidemiological models with standard incidence in Sect. 7.5.6.

Chapter 3
Spatio-Temporal Patterns
of Information Diffusion

Abstract In this chapter we examine spatio-temporal patterns of information diffusion in online social networks. We present relevant results from study of a Digg dataset. We analyze spatio-temporal patterns with friendship hops as distance in the Digg dataset. Finally, we discuss spatio-temporal patterns with shared interests as distance in the Digg dataset. For both distance metrics, spatio-temporal functions of influence exhibit S shape and are similar to logistic functions.

3.1 Introduction

Spatio-temporal models play a crucial role in the development of partial differential equation models in physics, mathematical biology, and other fields [74, 75], particularly, in describing the spread of infectious diseases [15]. In this chapter we analyze spatio-temporal patterns of the diffusion of news stories in Digg. Spatial models can provide a quantitative way to integrate local interactions between individual users into global views of information spread over social media. Part of the materials in this chapter are based on three published papers by the authors [128–130].

3.2 Digg Data Studies

The data used in this study were collected in a previous study [68], in which Lerman et al. collected the news stories that were promoted to the front page of digg.com, a major social news aggregate site, because of its popularity during June 2009. For each news story, the dataset includes a list of Digg

© Springer Nature Switzerland AG 2020

H. Wang et al., *Modeling Information Diffusion in Online Social Networks with Partial Differential Equations*, Surveys and Tutorials in the Applied Mathematical Sciences 7, https://doi.org/10.1007/978-3-030-38852-2_3

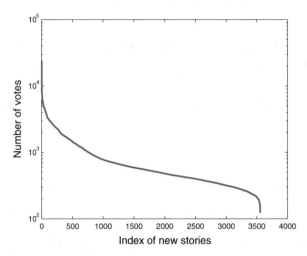

Fig. 3.1 Distribution of votes received by all the news studies in the datasets

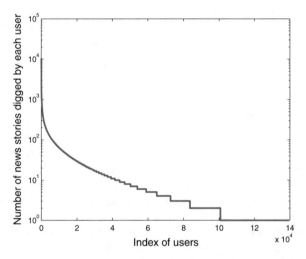

Fig. 3.2 Distribution of news stories digged by users

users who had *voted* on or *dugg* the story and the associated time-stamp. In total, for the most popular 3553 news stories during that month, 139,409 users cased more than three million votes (also called "diggs"). Figure 3.1 shows the distribution of the votes for all the news stories in the dataset. On average, each news story receives votes from nearly 850 users (0.6% of all the users in the dataset).

Figure 3.2 illustrates a *heavy-tail* distribution of news stories dugg by each user. As shown in this figure, there exist a few users who have actively voted on a large number of news stories, while over 70% of users have voted on 10 or less news stories.

3.3 Friendship Hops as Distance

3.3.1 Friendship Hops

In online social networks, cyber-distances of two users play a significant role in information diffusion. Cyber-distances are used to measure the closeness of users in online social networks. An intuitive approach for defining the distance between two users is to use the number of friendship links in the shortest path from one user to another in the social network graph, called *friendship hops.* Thus the distance between the initiator and any other user is defined as the length (the number of friendship hop) of the shortest path from the initiator to this user in the social network graph [128]. Clearly, the direct followers of the initiator have a distance of 1, while their own direct followers have a distance of 2 from the initiator, and so on. Figure 3.3 shows the distance distributions of the direct and indirect followers from Digg users who have initiated one or more top news stories in the Digg dataset. As we can see from the figure, the majority of online social network users have a distance of 2–5 from the initiators. In this figure, for all four randomly selected stories, the distance 3 users account for more than 40% of all the users from the initiator directly or through other users. As the distance increases from 6 to 8, the number of social networks users reachable from the initiator drops dramatically.

To be more precise, let U denote the user population in an online social network, and s be the source of information such as a news story that starts to spread in an online social network. Based on the distance from social network users from this source, the user population U can be divided into a set of groups; i.e., $U = \{U_1, U_2, \ldots U_i, \ldots, U_m\}$, where m is the maximum distance from the users to the source s. The group U_x consists of users who share the same distance of x to the source.

While the social distance defined in this chapter is based on friendship hops, the definition of cyber-distance can be flexible and may be defined as other measurements. For example, Sect. 3.4 discusses an alternative way to define distance metrics based on shared interests.

To model the information diffusion process in temporal and spatial dimensions, we examine the real datasets to analyze information spreading patterns over social media. With the definition of distance, all users can be divided into distance groups based on their distances from the news submitter. As a news story propagates through the Digg network, users express their in-

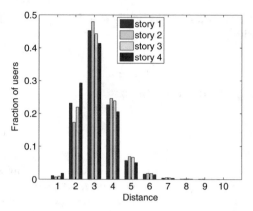

Fig. 3.3 Distribution of neighbors of four randomly selected stories

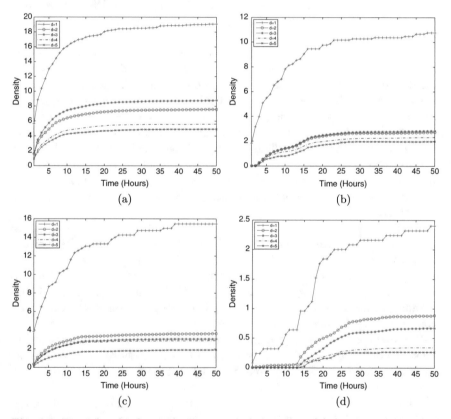

Fig. 3.4 Densities of influenced users over 50 h for s1–s4. (**a**) Density of influenced users of s1. (**b**) Density of influenced users of s2. (**c**) Density of influenced users of s3. (**d**) Density of influenced users of s4

terests in the news by voting or commenting on it. We denote such users as *influenced users* of the information.

3.3.2 Logistic Influence Curve

We studied the impact of friendship hop distance on the information diffusion process [128, 130] by measuring the *density of influenced users* within the same distance, which is the percentage of influenced users over the total number of users within a given distance. In the context of Digg social networks, we consider the users who have voted the news story as influenced users. Figure 3.4a–d illustrate the density of influenced users (with the distances of 1 to 5) over the initial 50 h after the news stories were posted on Digg for four example news stories, respectively. Each curve in Fig. 3.4a–d represents the density at a different distance.

We can observe from Fig. 3.4a–d that the densities of influenced users at different distances show consistent evolving patterns rather than increasing or decreasing with random fluctuations. The temporal and spatial patterns resemble dynamics of evolution equations involving both time and space variables. In addition, [130] validates the observations for all news stories in the Digg dataset. It is concluded that 94.9% of all news stories have the similar consistent evolving patterns. For most of the news stories, densities of influenced users decrease as the distances of the users increase, reconfirming that friendship is an important channel of information spreading. In addition, the news stories exhibit similar shapes of logistic curve with respect to time. Therefore, mathematical models, particularly, evolution equations involving both time and space variables, can be used to describe the evolution dynamics of information diffusion over social media.

3.4 Shared Interests as Distance

3.4.1 Shared Interests

In this section, we explore spatio-temporal patterns of information flow with shared interests. As online social networks continue to grow in users, traffic and information, it has become increasingly important to understand the interactions of people and information on these networks. For example, the discovery of shared interests among users could improve the quality of recommendation systems [70, 116, 143] or detect community of interests in social networks.

We believe that the insights gained from the discovery of shared interests among users will facilitate information spreading and the classifications on contents circulating in these networks, which will be very valuable for filtering unwanted contents such as spams in online social networks.

To gain an in-depth understanding of the interactions of users and information, we first study the shared interests of users on the news stories. A simple Digg activity or vote behavior $d_{u,s}$ suggests the interest of the user u on the news story s. Hence a pair of Digg activities, $(d_{u,s}, d_{v,s})$ indicates the *shared* or *common* interests of two users u and v on the same news stories s. Discovering such shared interests of all users in social news sites not only helps us to understand the interactions of users and information but also provides valuable insights for the classifications of users and information into certain categories.

The shared interests of two users u_i and u_j could be numerically represented with the number of the shared information they have interacted with, i.e., $s_{i,j} = \frac{|S_i \cap S_j|}{|S_i \cup S_j|}$, where S_i and S_j denote the set of information the users u_i and u_j have interacted with, respectively. For each news story, we first calculate the interest distance between the initiator and all other users, and classify the users into five disjoint groups based on five interest ranges, i.e., 0–0.2, 0.2–0.4, 0.4–0.6, 0.6–0.8, and 0.8–1. To make the distance values consistent with friendship hops, we use $d = 5$ to represent the first group, and $d = 4$ to represent the second group, and so on.

3.4.2 Logistic Influence Curve

Now we study the impact of the distance based on shared interests on the density of influenced users. Figure 3.5 shows the density of influenced users (with the interest distances of 1–5) over the initial 50 h for the same four news stories as those in Fig. 3.4, respectively. As show in Fig. 3.5, the density of influenced users is much higher for the users with a distance of 1 from the initiator than for those with a larger interest distance. For most of the news stories, densities of influenced users decrease as the distances of the users increase, reconfirming that shared interest is an important channel of information spreading. In addition, the news stories exhibit similar shapes of logistic curve with respect to time. Hence, the distance based on shared interest also serves as a good distance metrics for characterizing the process of information spreading over online social networks.

3.4.3 Studies of Shared Interests

In this section, we present a methodology to discover the clusters of users and information based on shared interest of users. Although there exists an extensive body of research work on network topology and on traffic characteristics and information spreading in online social networks [9, 50, 62, 85, 86, 111, 128], little attempt has been made to discover the clusters of users with shared interests or the clusters of information with which similar sets of social networks users interact.

With real datasets collected from Digg, a popular social news aggregation site, our experiment results show that the clustering step indeed divides the users and information into clusters with distinct characteristics. More importantly, the availability of the *users* or *information* clusters significantly improves our understanding of the interactions of users and information in online social networks.

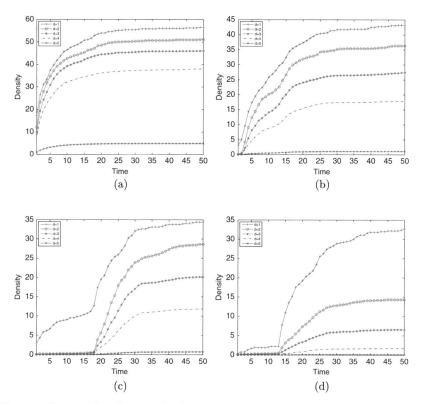

Fig. 3.5 Density distribution of infected users of a certain distance over 50 h with shared interest as distance. (**a**) Density distribution of story 1 with 24,099 votes. (**b**) Density distribution of story 2 with 8521 votes. (**c**) Density distribution of story 3 with 5988 votes. (**d**) Density distribution of story 4 with 1618 votes

In online social networks, the interactions of users and information could be naturally modeled with *bipartite graphs*, where users and information form two disjoint vertex sets [6, 37, 39, 110]. Bipartite graphs have been widely used in analyzing collaboration networks such as the co-authorship of authors on Wikipedia's articles and the collaborations of actors on movies [49, 91], since the graph structure of these collaboration networks exhibits bipartite patterns.

For the case of online social networks, we use bipartite graphs to characterize the interactions of users and news stories, and then build one-mode projection graphs on users and news stories, respectively. Further we run the clustering algorithm on the similarity matrix of the one-mode projection graphs and obtain clusters of users and news stories. The *users clusters* group users with shared interests on news stories, while the *news story clusters* group news stories that are voted by a similar set of users.

Specifically, we use a bipartite graph $G = \{U, S, E\}$ to represent the interactions between users and information. The vertex sets U and S denote all users and information in online social networks, respectively, while E denote the interactions between users and information. For example, $e_{u,s} \in E$ suggests that the user u interacts with the information s by creating, posting, commenting, digging, tweeting, or other activities in online social networks. Hence the $e_{u,s}$ reflects the interest of the user u on the information s. In other words, all the edges in the bipartite graph characterize the interactions between users and information in online social networks, and provide a unique perspective for understanding *why users interact with a certain set of information.*

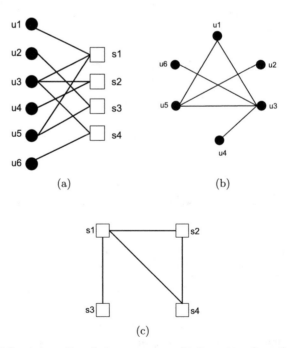

Fig. 3.6 Modeling interactions between users and information (tweet) with bipartite graphs and one-mode projection graphs. (**a**) An example of bipartite graphs. (**b**) One-mode projection graph on users. (**c**) One-mode projection graph on information

Based on the bipartite interaction graph of users and information in online social networks, we obtain two one-mode projection graphs: $G_U = \{U, E_U\}$ and $G_S = \{S, E_S\}$. An edge $e_{i,j}$ forms between two users u_i and u_j in G_U if and only if both of them interacts with one or more same information. The weighted edges of the graph G_U represent the degree of common interests. Similarly, two information nodes s_i and s_j are connected in G_S if and only if one or more users interact with both of them in the bipartite graph. The weighted edge between two information nodes represents the number of users that interact with both of them. Therefore, the one-mode projection graph G_U essentially captures the similarity of users based on their shared interests on information, while the graph G_S characterizes the similarity of information based on the user interaction patterns.

The edges of the one-mode projection graph have generated between two nodes in the same vertex set if both nodes connect to one or more same nodes in another vertex set of the bipartite graph, e.g., two users in online social networks tweeting on the same news story. As a result, one-mode projection graphs are often used to capture the hidden information or structure from the nodes in the same vertex set [37, 137].

Figure 3.6a–c illustrate an example of bipartite graphs and their one-mode projection graphs on both sets of vertices. Figure 3.6a shows six users in the left vertices set and four news stories in the right vertices set, while Fig. 3.6b, c show the one-mode projection graphs on the left and right vertex sets, respectively. The clique formed by nodes u_1, u_3, u_5 in Fig. 3.6b is due to their common connections to the same story s_1, while the clique of s_1, s_2, s_4 in Fig. 3.6c results from the user u_3 interacting with these news stories in the bipartite graph, as shown in Fig. 3.6a.

The edges in one-mode projection graphs serve as the *similarity* between the information or the *shared interests* between users. Thus, we further build the *similarity matrix* of the one-mode project graphs based on the connections among the information or users.

The availability of the similarity matrix leads us to the next step of finding clustering algorithms to discover clusters of users and information that share similar characteristics [52, 92, 136], because clustering algorithms have been widely used in clustering communication patterns of Internet end systems in recent years [43, 60, 134]. The similarity matrix in turn leads us to explore clustering algorithms to divide the users and information into distinctive and meaningful clusters such that each cluster groups users with similar interests or information with similar contents or topics. The goal of the clustering step is to divide the users and information into different groups based on their interaction patterns.

Here we adapt the agglomerative clustering algorithm on the similarity matrix of one-mode projection graphs G_U and G_S, since agglomerative algorithms optimize the clustering results by maximizing the internal similarity within the same clusters as well as minimizing the external similarity between nodes in different clusters [52]. We evaluate the quality of clustering results with the default I_2 criterion function [52], i.e., maximizing

$$\sum_k^{i=1} \sqrt{\sum_{v,u \in C_i} similarity(v,u)}, \tag{3.4.1}$$

where C_i denotes the i-th cluster, $i = 1, 2, \ldots, k$ and the $similarity(v, u)$ between information or users in the same clusters is computed with the $cosine$ function.

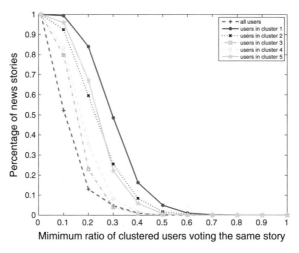

Fig. 3.7 Percentage of news stories that are voted on by users in the same *user clusters*

To evaluate the clustering results, we propose a simple metric, *voting consistency*, to denote the similarity of voting patterns in *users clusters* and *news story clusters*. Specifically, the voting consistency of a *user cluster* is measured by the percentage of news stories that are voted by a certain ratio of users in the same cluster. Figure 3.7 shows the percentage of news stories that are voted by users in the same clusters, where x-axis is the ratio of users in a given cluster and y-axis denotes the percentage of news stories that are voted by the ratio of users exceeding x in the same cluster. For example, in the *user cluster* 1 nearly 50% of news stories are voted on by at least 30% users in the same cluster. Such high voting consistency is significant and interesting, since compared with all the news stories, only 4.7% news stories receive votes from 30% of all users. In another instance, in the *user cluster* 5 more than 65% of news stories are voted on by at least 20% of all users in the cluster. However, only 12.9% news stories receive votes from more than 20% of all users. Similar observations on the high voting consistency hold in other *user clusters* as well.

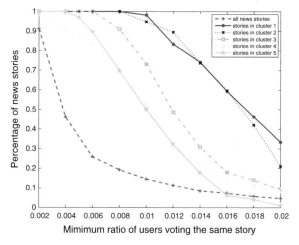

Fig. 3.8 Percentage of news stories that are voted by users in the same *news story clusters*

Next we evaluate the clustering results on the news stories with the same *voting consistency* metric. Like Figs. 3.7 and 3.8 illustrates the percentage of news stories that are voted on by a certain ratio of users for *news story clusters*. The overall observation is that the news stories in individual clusters receive consistent votes from similar groups of users compared with all the news stories (the red bottom line in the figure) in the entire dataset. These results suggest that the discovery of *news story clusters* successfully divides news stories into distinctive groups.

Chapter 4
Clustering of Online Social Network Graphs

Abstract In this chapter we briefly introduce graph models of online social networks and clustering of online social network graphs. We discuss graph models of online social networks and properties of Laplacian matrices. We focus on graph partitioning with eigenvectors of Laplacian matrices. We also present a clustering method based on higher-order organizations of graphs. Finally, we present spectral co-clustering with bipartite graphs.

4.1 Introduction

Online social networks can be conveniently modeled by graphs, which we often refer to as a social graph. The individuals within a network are the nodes, and an edge connects two nodes if the nodes are related by the relationship that characterizes the network. The explosive growth of social media in recent years has attracted millions of end users, thus creating social graphs with millions of nodes and billions of edges reflecting the interactions and relationship between these nodes.

Online social networks often exhibit community structure with inherent clusters. Detecting clusters or communities is one of the critical tasks in social network analysis because of its broad applications to matters such as friend recommendations, link predictions, and collaborative filtering. From the graph theory perspective, clustering and community detection essentially are to discover a group of nodes in a graph that are more connected with each other within the group than those nodes outside the group. Given the size and complexity of todays' online social networks, clustering and community detection in these networks face the inherent challenges.

For PDE modeling of information diffusion, communities (clusters) are essential to gain spatio-temporal inside into big datasets from online social

© Springer Nature Switzerland AG 2020

H. Wang et al., *Modeling Information Diffusion in Online Social Networks with Partial Differential Equations*, Surveys and Tutorials in the Applied Mathematical Sciences 7, https://doi.org/10.1007/978-3-030-38852-2_4

networks as we demonstrate from Chap. 3. Spatial distances in our PDE models often describe the strength of network connectivity among communities (clusters) rather than individual nodes. As a result, our PDE models are able to capture key characteristics of information spreading processes in an online social network.

In this chapter, we discuss graph models of online social networks and clustering techniques. Although there exist many clustering algorithms such as k-means clustering, in this book we focus on spectral clustering analysis because our data from online social networks are graph-structured. Part of the materials in this chapter is based on the authors' papers [132, 150] on spectral clustering as well as PDE modeling on online social networks.

4.2 Graph Models of Online Social Networks

In this section, we briefly review some of the common notation used in graphs. Any graph consists of both a set of objects, called nodes, and the connections between these nodes, called edges. Mathematically, a graph G is denoted as pair $G(V, E)$, where $V = \{v_1, v_2, \ldots v_n\}$ represents the set of nodes and $E = \{e_1, e_2, \ldots, e_m\}$ represents the set of edges and the size of the set is commonly shown as $m = |E|$. Edges are also represented by their end points (nodes), so $e(v_1, v_2)$ or (v_1, v_2) defines an edge between nodes v_1 and v_2. Edges can have directions if one node is connected to another, but not vice versa. When edges have directions, $e(v_1, v_2)$ is not the same as $e(v_2, v_1)$. When edges are undirected, nodes are connected both ways and are called *undirected edges* and this kind of graph is called *an undirected graph*. Graphs that only have directed edges are called *directed graphs* and ones that only have undirected edges are called *undirected graphs*. Finally, mixed graphs have both directed and undirected edges.

A sequence of edges where nodes and edges are distinct, $e_1(v_1, v_2)$, e_2 (v_2, v_3), $e_3(v_3, v_4), \ldots, e_i(v_i, v_{i+1})$, is called a path. A closed path is called a cycle. The length of a path or cycle is the number of edges traversed in the path or cycle. In a directed graph, we only count directed paths because traversal of edges is only allowed in the direction of the edges. For a connected graph, multiple paths can exist between any pair of nodes. Often, we are interested in the path that has the shortest length. This path is called the shortest path. We will also use the shortest path as distance for PDE modeling in online social networks. The concept of the neighborhood of a node v_i can be generalized using shortest paths. An n-hop neighborhood of node v_i is the set of nodes that are within n hops distance from the node v_i. We will use the n-hop neighborhood to cluster online users.

The degree of a node in a graph, which is the number of edges connected to the node, plays a significant role in the study of graphs. For a directed graph, there are two types of degrees (1) in-degrees (edges toward the node) and (2) out-degrees (edges away from the node). In a network, nodes with the most connections possess the greatest degree of centrality. Degree centrality measures relative levels of importance. We often regard people with many interpersonal connections to be more important than those with few. In-degree centrality describes the popularity of a node and its prominence or prestige. Out-degree centrality describes the gregariousness of the node. For social media, degree represents the number of friends for each given user. On Facebook, a degree represents the number of friends. For Twitter, in-degree and out-degree show the number of followers and followees, respectively.

A graph with n nodes can be represented by a $n \times n$ adjacency matrix. A value of 1 at row i, column j in the adjacency matrix indicates a connection between nodes v_i and v_j, and a value of 0 denotes no connection between the two nodes. When generalized, any real number can be used to show the strength of connection between two nodes. In directed graphs, we can have two edges between i and j (one from i to j and one from j to i), whereas in undirected graphs only one edge can exist. As a result, the adjacency matrix for directed graphs is not in general symmetric, whereas the adjacency matrix for undirected graphs is symmetric $(A = A^T)$. In social media, there are many directed and undirected networks. For instance, Facebook is an undirected network and Twitter is a directed network.

Consider a weighted graph $G = (V, E)$ with n vertices and m edges each with weights $E_{i,j}$ connecting nodes i, j. The adjacency of matrix M of a graph is defined by $M_{ij} = E_{ij}$ if there is an edge $\{i, j\}$ and $M_{ij} = 0$, otherwise. The Laplacian matrix L of G is an n by n symmetric matrix, with one row and column for each vertex, such that

$$
L_{ij} = \begin{cases} \sum_k E_{ik} & i = j \\ -E_{ij} & i \neq j, \text{and } v_i \text{ is adjacent to } v_j \\ 0 & \text{otherwise} \end{cases}
$$

In addition, a $n \times m$ incidence matrix of G, denoted by I_G has one row per vertex and one column per edge. The column corresponding to edge $\{i, j\}$ of I_G is zero except the i-th and j-th entries, which are $\sqrt{E_{ij}}$ and $-\sqrt{E_{ij}}$, respectively.

Now we have the following properties of L. Its proof can be found in many references (e.g., [21, 24]). Theorem 4.2.1 (7) is a result of the Rayleigh Quotient theorem (e.g., [45])

Theorem 4.2.1 The Laplacian matrix L has the following properties.

1. $L = D - M$, where M is the adjacency matrix and D is the diagonal degree matrix with $D_{ii} = \sum_k E_{ik}$.
2. $L = I_G I_G^T$.
3. L is symmetric positive semi-definite. All eigenvalues of L are real and nonnegative, and L has a full set of n real and orthogonal eigenvectors.
4. Let $e = [1, \ldots, 1]^T$. Then $Le = 0$. Thus 0 is the smallest eigenvalue and e is the corresponding eigenvector.

5. If the graph G has c connected components, then L has c eigenvalues that is 0.
6. For any vector x, $x^T L x = \sum_{\{i,j\} \in E} E_{ij}(x_i - x_j)^2$.
7. The problem

$$min_{x \neq 0} x^T L x, \text{ subject to }, x^T x = 1, x^T e = 0, \qquad (4.2.1)$$

is solved when x is the eigenvector corresponding to the second smallest eigenvalue (the Fiedler vector) λ_2 of the eigenvalue problem

$$L x = \lambda x. \qquad (4.2.2)$$

4.3 Spectral Graph Bipartitioning

For a weighted graph $G = (V, E)$, given a bipartition of V into disjoint V_1 and V_2 ($V_1 \cup V_2 = V$), the cut between them can be defined as

$$cut(V_1, V_2) = \sum_{i \in V_1, j \in V_2} M_{ij}. \qquad (4.3.1)$$

The definition of cut is easily extended to k vertex subsets

$$cut(V_1 V_2, \ldots . V_k) = \sum_{i < j} cut(V_i, V_j). \qquad (4.3.2)$$

The classical graph bipartitioning problem is to find nearly equally sized vertex subset V_i, V_2 of V such that $cut(V_1^*, V_2^*) = min_{V_1, V_2}$ cut (V_1, V_2). For this purpose, let us define the partition vector p that captures this division.

$$p_i = \{ \begin{matrix} +1, i \in V_1, \\ -1, i \in V_2. \end{matrix} \qquad (4.3.3)$$

The cut can be characterized by the Rayleigh Quotient as follows.

Theorem 4.3.1 Given the Laplacian matrix L of G and a partition vector p, the Rayleigh Quotient

$$\frac{p^T L p}{p^T p} = \frac{1}{n} \cdot 4cut(V_1, V_2). \qquad (4.3.4)$$

The result can be simply proved by the properties of the Laplacian matrix. The result indicates that the minimization of the cut can be represented by the Rayleigh Quotient with some partition vector (p_i) whose values are either -1 or 1. In practical applications, we also need an objective function to balance cuts. Such an objective function can be formulated as follows. Let us define a diagonal matrix with W where w_{ii} is a weight for each vertex i.

For a subset of vertices V_l, define its weight to be weight $W_{V_l} = \sum_{i \in V_l} W_{ii}$. Now we try to balance subsets V_1 and V_2 in such a way that the following objective function, $Q(V_1, V_2)$, is minimized.

$$Q(V_1, V_2) = \frac{cut(V_1, V_2)}{W_{V_1}} + \frac{cut(V_1, V_2)}{W_{V_2}}. \tag{4.3.5}$$

The minimization of $Q(V_1, V_2)$ favors partitions that have a small cut value and are balanced because for two different partitions with the same cut value, the above objective function value is smaller for the more balanced partitioning.

The objective function can be characterized by the Rayleigh Quotient of the following generalized partition vector q. Recall that all eigenvalues of L are real and nonnegative, and 0 is the smallest eigenvalue of L. For a given graph G, let L and W be its Laplacian and vertex weight matrices, respectively. Let $e = [1, \ldots, 1]^T$, $\nu_1 = W_{V_1}$, and $\nu_2 = W_{V_2}$, then the following result holds.

Theorem 4.3.2 The serialized partition vector $q = (q_i)$

$$q_i = \begin{cases} +\sqrt{\frac{\nu_2}{\nu_1}}, & i \in V_1, \\ -\sqrt{\frac{\nu_1}{\nu_2}}, & i \in V_2 \end{cases} \tag{4.3.6}$$

satisfies

1.
$$q^T W e = 0, \quad q^T W q = \nu_1 + \nu_2$$

2.
$$\frac{q^T L q}{q^T W q} = \frac{cut(V_1, V_2)}{\nu_1} + \frac{cut(V_1, V_2)}{\nu_2}, \tag{4.3.7}$$

3. The problem

$$min_{q \neq 0} \frac{q^T L q}{q^T W q}, \quad \text{subject to } q^T W e = 0, \tag{4.3.8}$$

is solved when q is the eigenvector corresponding to the second smallest eigenvalue λ_2 of the generalized eigenvalue problem,

$$Lx = \lambda W x. \tag{4.3.9}$$

Now we choose a weight$(i) = 1$ for all vertices i. This leads to the ratio-cut objective,

$$\text{Ratio-cut}(V_1, V_2) = \frac{cut(V_1, V_2)}{|V1|} + \frac{cut(V_1, V_2)}{|V_2|}. \tag{4.3.10}$$

One commonly used W is to choose W_{ii} to be the sum of the weights of edges incident on it, i.e., $W_{ii} = \sum_k E_{ik}$. This leads to the normalized cut criterion that was for image segmentation. Note that for this choice of vertex weights, the vertex weight matrix W equals the degree matrix D, and weight $(V_i) = cut(V_1, V_2) + \text{within}(V_i)$ for $i = 1, 2$, where within (V_i) is the sum of the weights of edges with both end points in V_i. Then the normalized-cut objective function may be expressed as

$$\text{Normalized cut}(V_1, V_2) = \frac{\text{cut}(V_1, V_2)}{\sum_{i \in V_1} \sum_k E_{ik}} + \frac{\text{cut}(V_1, V_2)}{\sum_{i \in V_2} \sum_k E_{ik}} = 2 - S(V_1, V_2),$$

(4.3.11)

where $S(V_1, V_2) = \frac{\text{within}(V_1)}{W_{V_1}} + \frac{\text{within}(V_2)}{W_{V_2}}$. Note that $S(V_1, V_2)$ describes the strengths of associations within each partition. As a result, minimizing the normalized cut is to maximize the proportion of edge-weights that lie within each partition while balancing the cut.

4.4 Clustering Based on Higher-Order Organization

Many network properties can be obtained through lower-order connectivity patterns that can be captured at the level of individual nodes and edges. In this section, we discuss higher-order organization of complex networks at the level of small network subgraphs which will help us reveal more insights of complex networks. We now briefly cover the background and theory for network clustering on the basis of higher-order connectivity patterns. Benson et al. [12] provide mathematical details on the optimality of obtained clusters and scales to networks with billions of edges.

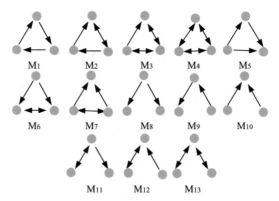

Fig. 4.1 Illustration of some common network motifs

We now define network motifs and more discussions can be found in [12]. We consider motifs to be a pattern of edges on a small number of nodes. Consider a weighted, undirected graph $G = (V, E)$, with $|V| = n$. Formally, we define a motif on k nodes by a tuple (B, A), where B is a $k \times k$ binary matrix and $A \subset \{1, 2, .., k\}$ is a set of anchor nodes. The matrix B encodes the edge pattern between the k nodes, and A labels a relevant subset of nodes for defining motif conductance. In many cases, A is the entire set of the nodes. Let $\mathcal{X}_A$ be a selection function that takes the subset of a k-tuple indexed by A, and let $set(\cdot)$ be the operator that takes an (ordered) tuple to an (unordered) set. Specifically,

$$set((v_1, v_2, \ldots, v_k)) = \{v_1, v_2, \ldots, v_k\}.$$

The set of motifs in an unweighted (possibly directed) graph with adjacency matrix A, denoted $M(B, A)$, is defined by

$$M(B, A) = \{(set(v), set(\mathcal{X}_A(v))) | v \in V^k, v_1, v_2, \ldots, v_k, distinct, A_v = B\},$$

where $A_v = B$ is the $k \times k$ adjacency matrix on the subgraph induced by the k nodes of the ordered vector v. We call any $(v, \mathcal{X}_A(v)) \in M(B, A)$ a motif instance. When B and A are arbitrary or clear from context, we will simply denote the motif set by M. We call motifs where $\mathcal{X}_A(v) = v$ simple motifs and motifs where $\mathcal{X}_A(v) \neq v$ anchored motifs. Motif analysis in the literature has mostly analyzed simple motifs. However, the anchored motif provides us with a more general framework (Fig. 4.1).

We now define motif conductance. We now relate these matrices to the conductance of a set S with $\overline{S} = V \setminus S$. Recall that the key definitions for conductance are the notions of cut and volume. Informally, $cut_M^{(G)}(S, \overline{S})$ and $vol_M^{(G)}(S)$ can be defined as the number of motif instances in the cut and the number of instance end points in S. Formally, we define motif conductance as

$$\phi_M^{(G)}(S) = cut_M^{(G)}(S, \overline{S}) / \min(vol_M^{(G)}(S), vol_M^{(G)}(\overline{S})),$$

where

$$cut_M^{(G)}(S, \overline{S}) = \sum_{(v, \mathcal{X}_A(v)) \in M} \mathbf{1}(\exists i, j \in \mathcal{X}_A(v) | i \in S, j \in \overline{S})$$

$$vol_M^{(G)}(S) = \sum_{(v, \mathcal{X}_A(v)) \in M} \sum_{(i \in \mathcal{X}_A(v))} \mathbf{1}(i \in S)$$

and $\mathbf{1}(s)$ is the truth-value indicator function on s, i.e., $\mathbf{1}(s)$ takes the value of 1 if the statement s is true or 0 otherwise.

We say that a motif instance is cut if there is at least one anchor node in S and at least one anchor node in $\overline{S}$. The motif cut measure counts an instance of a motif as cut only if the anchor nodes are separated, and the motif volume counts the number of anchored nodes in the set. However, two nodes in an anchor set may a part of several motif instances. Specifically, in the above definition, there may be many different v with the same $\mathcal{X}_A(v)$, and the nodes in $\mathcal{X}_A(v)$ still get counted proportional to the number of motif instances.

Now we define motif adjacency matrix and motif Laplacian. Given an unweighted, directed graph and a motif set M, we conceptually define the motif adjacency matrix by $(W_M)_{ij}$ as the number of motif instances in M where i and j participate in the motif. Or, formally,

$$(W_M)_{ij} = \sum_{(v,\mathcal{X}_A(v)) \in M} \mathbf{1}(\{i,j\} \subset \mathcal{X}_A(v))$$

for $i \neq j$. Note that weight is added to $(W_M)_{ij}$ only if i and j appear in the anchor set. Next, we define the motif diagonal degree matrix by $(D_M)_{ii} = \sum_{j=1}^{N}(W_M)_{ij}$ and the motif Laplacian as $L_M = D_M - W_M$. Finally, the normalized motif Laplacian is $\mathcal{L}_M = D_M^{-1/2}(D_M - W_M)D_M^{-1/2} = I - D_M^{-1/2}W_M D_M^{-1/2}$, which results from the minimization of the normalized cut for a graph.

The following result states motif conductance on the original graph G is equivalent to conductance on the weighted graph G_M when there are three anchor nodes. In other words, when the number of anchor nodes is 3, the motif conductance is equal to the conductance on the weighted graph.

Theorem 4.4.1 Let $G = (V, E)$ be a directed, unweighted graph and let W_M be the weighted adjacency matrix for any motif with $|A| = 3$. Then for any $S \subset V$

$$vol_M^{(G)}(S) = vol_M^{(G_M)}(S).$$

We are now ready to describe the motif-based clustering algorithm for finding a single cluster in a graph as follows.

1. *Given* a directed, unweighted graph G and motif M
2. *Compute* motif-based clusters (subset of nodes in G)
   $(W_M)_{ij} \leftarrow$ number of instances of M that contain nodes i and j.
   $G_M \leftarrow$ weighted graph induced W_M.
   $D_M \leftarrow$ diagonal matrix with $(D_M)_{ii} = \sum_{j=1}(W_M)_{ij}$.
   $z \leftarrow$ eigenvector of second smallest eigenvalue for $\mathcal{L}_M = I - D_M^{-1/2}W_M D_M^{-1/2}$.
   $\sigma_i \leftarrow$ to be index of $D_M^{-1/2}z$ with ith smallest value.
   /* Sweep over all prefixes of σ */
   $S \leftarrow argmin_l \phi^{G_M}(S_l)$, where $S_l = \{\sigma_1, \ldots, \sigma_l\}$.

3. If $|S| < |\overline{S}|$, then return S else return $\overline{S}$.

The algorithm finds a partition of the nodes into S and $\overline{S}$. The motif conductance is symmetric in the sense that $\phi_G(S) = \phi_G(\overline{S})$ so either set of nodes (S or $\overline{S}$) could be interpreted as a cluster. The algorithm is based on

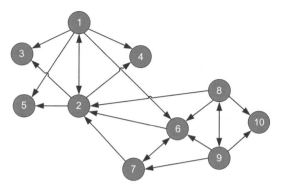

Fig. 4.2 Example graph from [12]

the Fiedler partition of the motif weighted adjacency matrix. It is shown in [12] that when the motif M has three nodes the above clustering algorithm is nearly optimal.

As an example, we consider a simply network graph in Fig. 4.2 to search for clusters ([150] also discusses another spectral partition example). For this graph and a motif of interest (in this case, M_7 is selected in Fig. 4.1), we can compute the weighted adjacent matrix is

$$W_M = \begin{pmatrix} 0 & 3 & 1 & 1 & 1 & 0 & 0 & 0 & 0 & 0 \\ 3 & 0 & 1 & 1 & 1 & 1 & 1 & 0 & 0 & 0 \\ 1 & 1 & 0 & 0 & 0 & 0 & 0 & 0 & 0 & 0 \\ 1 & 1 & 0 & 0 & 0 & 0 & 0 & 0 & 0 & 0 \\ 1 & 1 & 0 & 0 & 0 & 0 & 0 & 0 & 0 & 0 \\ 0 & 1 & 0 & 0 & 0 & 0 & 1 & 1 & 1 & 0 \\ 0 & 1 & 0 & 0 & 0 & 1 & 0 & 0 & 0 & 0 \\ 0 & 0 & 0 & 0 & 0 & 1 & 0 & 0 & 2 & 1 \\ 0 & 0 & 0 & 0 & 0 & 1 & 0 & 2 & 0 & 1 \\ 0 & 0 & 0 & 0 & 0 & 0 & 0 & 1 & 1 & 0 \end{pmatrix} \qquad (4.4.1)$$

and motif diagonal degree matrix $D_M = \text{dial}(6, 8, 2, 2, 2, 4, 2, 4, 4, 2)$ and corresponding motif Laplacian

$$L_M = D_M - W_M = \begin{pmatrix} 6 & -3 & -1 & -1 & -1 & 0 & 0 & 0 & 0 & 0 \\ -3 & 8 & -1 & -1 & -1 & -1 & -1 & 0 & 0 & 0 \\ -1 & -1 & 2 & 0 & 0 & 0 & 0 & 0 & 0 & 0 \\ -1 & -1 & 0 & 2 & 0 & 0 & 0 & 0 & 0 & 0 \\ -1 & -1 & 0 & 0 & 2 & 0 & 0 & 0 & 0 & 0 \\ 0 & -1 & 0 & 0 & 0 & 4 & -1 & -1 & -1 & 0 \\ 0 & -1 & 0 & 0 & 0 & -1 & 2 & 0 & 0 & 0 \\ 0 & 0 & 0 & 0 & 0 & -1 & 0 & 4 & -2 & -1 \\ 0 & 0 & 0 & 0 & 0 & -1 & 0 & -2 & 4 & -1 \\ 0 & 0 & 0 & 0 & 0 & 0 & 0 & -1 & -1 & 2 \end{pmatrix}. \qquad (4.4.2)$$

Following the above algorithm we sort nodes by values in the eigenvector of second smallest eigenvalue: $f_1, f_2, \ldots f_{10}$. Let $S_r = \{f_1, \ldots, f_r\}$ and compute the motif conductance of each S_r. It turns out that the motif conductance reaches its smallest number when $r = 5$. As a result, the network is divided into two clusters: one with nodes $1, 2, 3, 4, 5$ and one with the rest of nodes.

The above discussion is based on unweighted networks, where only the existence of connections between nodes is considered, and not their strength or capacity. In many real-world networks, edges contain more information than just simple node connectivity. To greatly improve the expressiveness and information content of the connections, in this section we incorporate weight information in the edges and the higher-order network analysis.

We extend the conductance of unweighted network to weighted network and thereby improve on the motif-based higher-order spectral clustering method in [12]. For weighted motifs, a formal expression was given, considering the weight of a motif instance. But how to define the weight for each motif instance is still an open question. In [132], the authors introduced a weighted support measure based on the weight of each motif instance for a weighted, directed graph G.

It is important to choose an appropriate measure to define the weight $\omega_{(v,\chi_A(v))}$ for every motif instance $(v, \chi_A(v))$ of motif M. We consider several situations to describe the weight of the motif instance. A simple method of calculating the weight of the motif instance is to use maximum weight of the edges or minimum weight of the edges as $\omega_{(v,\chi_A(v))}$. But these two measures cannot describe the motif instance as a whole. For example, two motif instances, one with a big-weight edge and a small-weight edge and the other with two big-weight edges, may well be assigned the same value. Another measure is to define the arithmetic mean of the edges' weights as the weight of the motif instance. However, this method does not consider differences between edges' weights, and it is not robust against extreme values of edges' weight. Thus we introduce a new approach like the geometric mean of the weights of edges as

$$\omega_{(v,\chi_A(v))} = \left(\prod_{(i,j) \in \chi_A(v)} \omega_{ij} \right)^d.$$

This measure can overcome some of the sensitivity issues of the arithmetic mean. If a motif instance is made up of an edge with a low weight value and an edge with a high weight value, the geometric mean will have a lower value.

As a result of this new measurement for the weights of edges, we further introduce the conductance of a node set S for a directed and weighted graph G as

$$\phi_M^G(S) = cut_M^G(S, \bar{S})/min(vol_M^G(S), vol_M^G(\bar{S})),$$

where

$$cut_M^G(S, \bar{S}) = \sum_{(v, \chi_A(v)) \in M} \left(\prod_{(i,j) \in \chi_A(v)} \omega_{ij} \right)^d \mathbf{1}\left(\{i, j\} \in \chi_A(v), i \in S, j \in \bar{S}\right),$$

$$vol_M^G(S) = \sum_{(v, \chi_A(v)) \in M} \left(\prod_{(i,j) \in \chi_A(v)} \omega_{ij} \right)^d \sum_{i \in \chi_A(v)} \mathbf{1}\left(i \in S\right),$$

$\mathbf{1}(s)$ takes the value of 1 if the statement s is true, or 0 otherwise, $\bar{S}$ denotes the remainder of the nodes (the complement of S). This definition of the conductance $\phi_M^G(S)$ is inspired by [12]. d is a threshold value, which can be adjusted in different conditions. $\left(\prod_{(i,j) \in \chi_A(v)} \omega_{ij}\right)^d \equiv \omega_{(v, \chi_A(v))}$ is the weight of the corresponding motif instance $(v, \chi_A(v))$, which shifts weights of discrete direct edges to the weight of a continuum of the subgraphs. Higher-order clustering algorithm aims to find a cluster (defined by S) that minimizes $\phi_M^G(S)$, with given motif M.

The definition of motif weight $\omega_{(v, \chi_A(v))}$ and the choice of d are illustrated for motif M_8 in Fig. 4.1. Motif M_8 reflects the movements from source to target. For motif M_8, weight of the motif instance $(v, \chi_A(v))$ is

$$w_{(v, \chi_A(v))} = \left(\prod_{(i,j) \in \chi_A(v)} \omega_{ij} \right)^d = (\omega_{12}\omega_{13})^d,$$

where ω_{12} is the weight of the directional edge from node 1 to node 2 just as M_8 in Fig. 4.1; the definition of ω_{13} is similar to ω_{12}, and d is a positive value. In [132], we apply $d = 1/2, 1, 2, \ldots$; experimental results demonstrate that the motif conductance ϕ decreases as d increases, and ϕ remains nearly stable when d arrives at 4 or even a larger number. This is in accordance with [12], which illustrated that lower conductance can lead to better clustering.

4.5 Spectral Partitioning with Bipartite Graph

The interactions between users and tags (information) can be represented by a undirected bipartite graph $G = (D, U, E)$ where $D = \{d_1, \ldots d_n\}$, $U = \{u_1, \ldots, u_m\}$ are two sets of vertices and E is the set of edges $\{\{d_i, u_j\} : d_i \in D, u_j \in U\}$. In the user-tag case, U is the set of users and D is the set of tags that the users are associated with. An edge $\{d_i, u_j\}$ exists if a user u_j associates with tag d_i. Note that edges are undirected. In this model there are no edges between users or tags. An edge represents a connection between a user and a tag. We can capture the strength of this association with positive weights on the edges. This also allows us to have edge-weights to represent the frequency with which a tag is associated with a user. In fact the bipartite graph in our flu prediction application in Sect. 8.3 has edge-weights to describe the frequency of a state is mentioned in tweets from a user.

Define $m \times n$ user-by-tag matrix A such that A_{ij} equals the edge-weights E_{ij}. It is easy to verify that the adjacency matrix of the bipartite graph may be written as

$$M = \begin{bmatrix} 0 & A \\ A^T & 0 \end{bmatrix},$$

where the vertices are ordered so that the first m vertices are the index of the users, while the last n index the tags. Because of the association of user and tag clustering in the bipartite graph, tag clustering will induce user clustering, while user clustering will induce tag clustering. Note that this procedure is recursive in nature since tag clusters determine user clusters, which in turn lead to natural user clusters. Clearly the optimal tag and user clustering would have the minimum weights for the crossing edges between partitions.

For normalized cut, it is shown that the second eigenvector of the generalized eigenvalue problem

$$Lz = \lambda Dz$$

gives a solution to find the minimum normalized cut. Here we present algorithms to find user and tag clusterings using our bipartite graph model. In the bipartite case,

$$L = \begin{pmatrix} D_1 & -A \\ -A^T & D_2 \end{pmatrix} \text{ and } D = \begin{pmatrix} D_1 & 0 \\ 0 & D_2 \end{pmatrix},$$

where D_1 $(m \times m)$ and D_2 $(m \times m)$ are diagonal matrices such that $D_1(i, i) = \sum_j A_{ij}, D_2(j, j) = \sum_i A_{ij}$. Thus $Lz = \lambda Dz$ may be written as

$$\begin{pmatrix} D_1 & -A \\ -A^T & D_2 \end{pmatrix} \begin{pmatrix} x \\ y \end{pmatrix} = \lambda \begin{pmatrix} D_1 & 0 \\ 0 & D_2 \end{pmatrix} \begin{pmatrix} x \\ y \end{pmatrix}. \tag{4.5.1}$$

Assuming that both D_1 and D_2 are nonsingular, the above equations can be rewritten as

$$\begin{aligned} D_1^{1/2} x - D_1^{-1/2} A y &= \lambda D_1^{1/2} x, \\ -D_2^{-1/2} A^T x + D_2^{1/2} y &= \lambda D_2^{1/2} y. \end{aligned} \tag{4.5.2}$$

Letting $u = D_1^{1/2} x$ and $v = D_2^{1/2} y$, we have

$$\begin{aligned} D_1^{-1/2} A D_2^{-1/2} v &= (1 - \lambda) u, \\ -D_2^{-1/2} A^T D_1^{-1/2} u &= (1 - \lambda) v, \end{aligned} \tag{4.5.3}$$

which can be solved by the singular value decomposition (SVD) of the normalized matrix

$$B = D_1^{-1/2} A D_2^{-1/2}.$$

In particular, u and v are the left and right singular vectors, respectively, while $(1 - \lambda)$ is the corresponding singular value. Thus in order to compute the eigenvector of the second (smallest) eigenvalue of (4.5.1), we only need to compute the left and right singular vectors corresponding to the second (largest) singular value of B

$$Bv_2 = \sigma_2 u_2, \quad B^T u_2 = \sigma_2 v_2, \tag{4.5.4}$$

where $\sigma_2 = 1 - \lambda_2$. From the computational point of view, we prefer to work on B instead of L because B is of size $m \times n$, while the matrix L is of the larger size $(n+m) \times (n+m)$. The right singular vector v_2 $(n \times n)$ will give us a bipartitioning of tags, while the left singular vector u_2 $((m \times m))$ will give us a bipartitioning of users. Equation (4.5.4) further indicates that a partitioning of tags should induce a partitioning of users, while a partitioning of users should imply a partitioning of tags.

The singular vectors u_2 and v_2 of B give a real approximation for the minimization of the normalized cut. Thus we now look for a bi-modal distribution in the values of u_2 and v_2. Let m_1 and m_2 denote the bi-modal values that we are looking for. From the previous section, the second eigenvector of L is given by

$$z_2 = \begin{pmatrix} D_1^{-1/2} \, u_2 \\ -D_2^{-1/2} \, v_2 \end{pmatrix}. \tag{4.5.5}$$

Now the classical k-means algorithm can be used to approximate the optimal bipartitioning such that the following sum of squares criterion is minimized,

$$\sum_{j=1}^{2} \sum_{z_2 \in m_j} (z_2(i) - m_j)^2. \tag{4.5.6}$$

Bipartition Algorithm

1. Given A, form $B = D_1^{-1/2} A D_2^{-1/2}$.
2. Compute the second singular vectors of B, u_2 and v_2 and form the vector z_2 as in (4.5.5).
3. Run the k-means algorithm on the 1-dimensional data z_2 to obtain the desired bipartitioning.

The above bipartitioning algorithm can be adapted for the more general problem of finding k tag and user clusters. To find k clusters, let $l = \lceil log_2 k \rceil$ and take singular vectors $u_2, u_3, \ldots u_{l+1}$, and $v_2, v_3, \ldots . v_{l+1}$. Thus we can form the l-dimensional dataset

$$Z = \begin{pmatrix} D_1^{-1/2} \, U \\ D_2^{-1/2} \, V \end{pmatrix}, \tag{4.5.7}$$

where $U = [u_2, \ldots . . u_{l+1}]$, and $V = [v_2, \ldots . . v_{l+1}]$. From this lower dimensional dataset, we look for the best k-modal fit to the l-dimensional points $m_1, \ldots, m_k$ by assigning each l-dimensional row, $Z(i)$, to m_j such that the sum of squares

$$\sum_{j=1}^{k} \sum_{Z(i) \in m_j} \|Z(i) - m_j\|^2 \qquad (4.5.8)$$

is minimized. This can again be done by the classical k-means algorithm. Thus we obtain the following algorithm.

Multipartition Algorithm

1. Given A, form $A_n = D_1^{-1/2} A D_2^{-1/2}$.
2. Compute $l = [log_2 k]$ singular vectors of $A_n, u_2 \ldots u_{l+1}$ and $v_2, \ldots v_{l+1}$, and form the matrix Z as in (4.5.7).
3. Run the k-means algorithm on the l-dimensional data Z to obtain the desired k-way multipartitioning.

4.6 Discussions

Major social media include news aggregation (Digg, Google, Yahoo, Baidu), social networking (Facebook or LinkedIn), microblogging (Twitter, Weibo), photo sharing (Flickr, Photobucket), video sharing (YouTube), and many others. Social media, taken together, have become a major platform for people to communicate and exchange information. Social media enable individuals and organizations to interact with one another through friendships, emails, blogposts, interest sharing, and many other mechanisms. Online social networks allow people to interact and communicate with each other, and to generate and disseminate information and content in the forms of texts, videos, and pictures. Complexity and diversity of entities and interactions in online social networks lead to a variety of dimensions and features in the datasets collected from online social networks.

Nodes in the same communities or clusters of social networks often exhibit strong connections, e.g., friendships, following/follower relationships, and share common interests in certain topics. Communities tend to have high clustering coefficient, which quantifies the closeness of the neighborhood of a node (let us name this node A) by measuring the number of connections between A's friends over the maximum possible connections among them. Two of A's "disconnected" neighbors in the community are likely to become friends if they are introduced or recommended to each other via A—their mutual friend. Communities (clusters) are one of the most important characteristics of online social networks, often revealing underlying structures of online social networks, and providing insight into the demographic identification of network components and the function of dynamical processes that operate on networks.

In this book we focus on spectral clustering as our data are graph-structured. Graph partitioning plays an important role in various applications, such as network design, circuit partitioning, telephone, load balancing in parallel computation, etc. However it is well known that this problem is NP-complete. There are a number of effective clustering algorithms. For example, modularity optimization [89] and k-clique percolation [97]. Barbieri et al. [7] introduces a community-cascade network model to characterize community structures and information diffusion of social networks in parallel, detects overlapping communities, and explores the discovered communities to understand information cascade and social contagion. Similarly, [72] tackles the problems of community detection and network discovery at the same time, such that these two mutually dependent processes could effectively support and enhance each other.

The proliferation of community detection and clustering algorithms allow researchers to explore a variety of algorithms to discover community and clusters from online social networks. However, selecting the best algorithm for each specific application is not an easy task. Thus a systematic evaluation method with meaningful and verifiable metrics is desired for determining the optimal algorithm. The availability of large scale data in online social networks has driven the recent effort on community detection to focus on efficient community detection using graph simplification algorithms to explore both link and content similarity to identify clusters and communities in large scale online social networks [6, 7, 25, 35, 37, 39, 41, 49, 52, 60, 62, 72, 92, 95, 97, 107, 113, 120, 134, 136, 137, 139]. In this chapter, we briefly introduce some basic knowledge of clustering of user and information and focus on spectral clustering methods using eigenvectors of matrices derived from data.

Chapter 5
Partial Differential Equation Models

Abstract In this chapter we present a number of partial differential equation models for information diffusion in online social networks with friendship hops as distance metrics. First, we discuss embedding of graphs of online social networks into Euclidean spaces. We develop a conceptual framework that divides the information diffusion process over online social networks into two separate processes: external and internal influences. We establish a general framework for partial differential equation models and discuss three diffusive logistic models and then validate them with the Digg and Twitter datasets.

5.1 Introduction

An important observation from the study of the Digg datasets is that the diffusion processes for different news stories exhibit similar temporal and spatial patterns of the growth of influence users. The temporal and spatial characteristics of information diffusion process shed light on how information spreads across different clusters over online social networks.

In this chapter, we will present a theoretical framework for spatio-temporal modeling of information diffusion in online social networks. The framework integrates graph clustering and embedding into Euclidean spaces with PDE modeling of information diffusion. We develop a conceptual framework that divides the information diffusion process in online social networks into two separate processes: external and internal influences. We will discuss three reaction-diffusion equation models based on the friendship hop distance and validate the models with Digg and Twitter datasets. Part of the materials in this chapter is based on the authors' papers [128, 130, 147, 148].

© Springer Nature Switzerland AG 2020

H. Wang et al., *Modeling Information Diffusion in Online Social Networks with Partial Differential Equations*, Surveys and Tutorials in the Applied Mathematical Sciences 7, https://doi.org/10.1007/978-3-030-38852-2_5

5.2 Embedding Network Graphs to Euclidean Spaces

In order to use partial differential equations to model information diffusion in online social network, it is essential to embed the corresponding graph into Euclidean spaces. For a given graph $G(V, E)$, Theorem 4.2.1 (6) includes the following relation:

$$x^T L x = \sum_{\{i,j\} \in E} E_{ij}(x_i - x_j)^2,$$

where x is a vector, L is the Laplacian matrix, and E_{ij} is the weights connecting nodes i, j. If we interpret $x = (x_i)$ as positions in a line, then Theorem 4.2.1 (7) indicates that the Fiedler vector (the eigenvector corresponding to the second largest eigenvalue), which is a particular (x_i) could be used to map a weighted graph onto a line such that connected nodes stay as close as possible [13]. For a small graph embedding of nodes of a graph may be acceptable. However, for a graph with larger number of nodes, we need to apply clustering algorithms to find meaningful communities/clusters. In this situation, we can treat clusters as a node in a new graph where the strength of edges are the summation of all weights between two clusters, then this approach can be used for embedding clusters into Euclidean spaces as illustrated in Fig. 5.1.

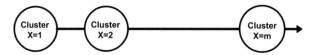

Fig. 5.1 Embedding of clusters into the x-axis

In addition, we could choose other meaningful methods relevant to a specific clustering technique. Assume that we breakdown the user population U into a set of groups, i.e., $U = \{U_1, U_2, \ldots U_i, \ldots, U_m\}$. If the clustering is based on the shortest friendship hops from the source, the group U_x consists of users who share the same distance of x to the source and m is the maximum distance from the users to the source s. For the friendship hop partition, as a result, there is a natural spatial arrangement of these clusters. We arrange U_i by the distance from the source. For partition based on shared interest, we arrange U_i by their shared interests. For general clustering partitions, the spatial arrangement of U_i can be based on specific modeling goals and social or geographical characteristics of the underlying network. In [63, 64] we use the level of democracy, diaspora size, international economic relations, or geographical proximity to order U_i. Here we choose the increment between each cluster to be 1, but it can be adjusted according to specific cases. Following [128], we use the x-axis as the social distance and embed the density U_x at location x.

If we intend to examine information diffusion along multiple social factors or communication channels, we may embed the graph into a high dimensional space R^n, as illustrated in Fig. 5.2. For higher dimensional mapping, we use more eigenvectors of the Laplacian matrix to minimize $x^T L x$ to make sure that connected nodes stay as close as possible. Section 6.6 briefly discusses partial differential equations defined on high dimensional spaces. However, we mainly focus on one-dimensional space in this book.

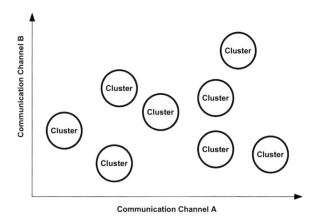

Fig. 5.2 Embedding in high dimensional Euclidean space with respect to two communication channels A and B

5.3 External and Internal Influences

In this section, we develop a conceptual framework that divides the information diffusion process in online social networks into two separate processes: external and internal influences. We will see that these two processes have different spatial effects on information diffusion as the role of clusters for information diffusion in each process differs.

In general, two decisive components for information diffusion in online social networks are the graph structure of social networks (follower graphs) and the content of the information, which form the backbone of online social networks [106]. Online users are subject to information from a wide range of sources, not just those networks they are connected to [123]. In the setting of online social networks, because each cluster consists of users with the similar social distance from a source, the growth of the influenced users within the group may be viewed as a result of the network structure. Other activities to promote information diffusion such as search may not be directly related

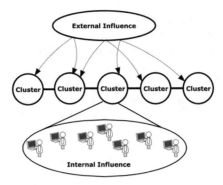

Fig. 5.3 External and internal influences

to the network structure and may happen randomly for various reasons, in most cases, mainly because of the information content.

Therefore, we divide the information diffusion process in online social networks into two separate processes: external and internal influences. In online social networks, the two types of influences resemble the external and internal influences, respectively, in [84]. The internal influences represent the information spreading among users within similar distances because of their direct links to those who are already influenced, and of course, the contents of information as well. The internal influences may result both from the network's structure and from the information content. The external influences measure information spread among users at different distance due to various other activities that result from the content of information. The external influences can be bidirectional or reciprocal in a manner of a random walk. The interplay of the two processes essentially accounts for the change of the density of influenced users. Figure 5.3 illustrates the conceptional interplay of the two processes in an online social network and their spatial effects on information diffusion.

The two influence framework directly addresses a number of concerns with diffusion models derived from ordinary differential equation models. Tufekci et al. [123] observed that there are significant differences between information traveling in social media and the spreading of germs in that online users are exposed to information from a wide range of sources and not only from the networks they are connected to. The same issue also was raised by Myers et al. [84] (also see [38]) where two different diffusion processes, internal and external influence, are discussed. The internal influence results from the structure of the underlying network; the external influence comes from various out-of-network sources, such as the mainstream media. While the network of social relationships is a major factor for diffusion of information, the popularity of the content of information is the key driving force behind the external influence. It is estimated in [84] that almost 27% of information volume in Twitter can be attributed to the external influence. Myers et al. [84] noticed that nearly all epidemiological models for online social networks only focus on the internal influence, while neglecting the external influence. However, the probabilistic model in [84] primarily focuses on separating the external influence from the internal influence, and quantifying the effect of the external influences on information adoption over time. The PDE-based models we developed integrate the effect of both internal influence and external influence through dynamical systems in both temporal and spatial dimensions. In this context, external and internal influences provide a new analytic framework for a better understanding of information diffusion mechanisms by studying the interplay of structural and topical influences.

5.4 PDE Model Formulation

Once we embed the information propagation process into Euclidean spaces, the formulation of the spatio-temporal model for information flow is similar to that for spatial biology [82]. Spatio-temporal models are based on a mathematical formulation of the basic fact that the rate at which a given quantity changes in a given domain must equal the rate at which it flows across its boundary plus the rate at which it is created, or destroyed, within the domain. We emphasize differences and their interpretations in social media.

In social media, the quantity is the amount of information spreading such as the density of influenced users, denoted by $I = I(x, t)$; and this is measured in amount per unit length along the x-axis since we are embedding the users of an entire network into a one-dimensional space. We assume that any change in the amount of information be restricted to one spatial dimensional tube where each cross-section is labeled as the spatial variable x. While only discrete set of points (U_x) in the x-axis is meaningful for social media, we can extend the discrete points into a continuous interval. With this understanding, we can develop the general spatio-temporal model for information flow, which further derives various partial differential models.

Because social media have rapidly gained worldwide popularity in recent years, many social media sites have experienced explosive growth in registered online users. For example, in 2019, Twitter has 1.5 billion registered users. This gives rise to extremely complex and large network graphs in online social networks. If we introduce a slightly more complex distance metric from the network's underlying topology, the number of the subsets in U can increase dramatically. Therefore the user population U will be embedded to some interval on the x-axis with more dense points. In particular, when we discuss traveling wave solutions, it is assumed that these discrete points are sufficiently dense on $(-\infty, \infty)$.

For simplicity, we assume that a constant A is the cross-sectional area of the tube. Thus the amount of information in a small section of width dx is $I(x, t)A dx$. Further, we let $J = J(x, t)$ denote the flux of the quantity at x, at time t. The flux measures the amount of the quantity crossing the section at x at time t, and its units are given in amount per unit area, per unit time. By convention, flux is positive if the flow is to the right, and negative if the flow is to the left.

Let $f = f(I, x, t)$ denote the given rate at which the information is created within the section at x at time t. Here, f represents the process within clusters in Fig. 5.2 and is a result of local growth often directly related to the underlying network structure. The process within clusters has much in common with the internal influence in [84]. In social media, f can be negative due to actions involving deletion; f is measured in amount per unit volume, per unit time. In this way, $f(I, x, t)Adx$ represents the amount of information that is created in a small width dx per unit.

We now can formulate the spatio-temporal model by considering a fixed, but arbitrary, section $a \leq x \leq b$ of the domain. The rate of change of the total amount of the information in the section must equal the rate at which it flows in at $x = a$, minus the rate at which it flows out at $x = b$, plus the rate at which it is created within $a < x < b$. In mathematical formulation, for any section $a \leq x \leq b$,

$$\frac{d}{dt} \int_a^b I(x,t)Adx = AJ(a,t) - AJ(b,t) + \int_a^b f(I,x,t)Adx.$$

From the fundamental theorem of calculus, $J(a,t) - J(b,t) = -\int_a^b \frac{\partial J}{\partial x}dx$. Because A is constant, it may be canceled from the formula. We arrive at, for any section $a \leq x \leq b$,

$$\int_a^b \left(\frac{\partial I}{\partial t} + \frac{\partial J}{\partial x} - f(I,x,t)\right)dx = 0.$$

It follows that the process of information diffusion can be formulated as

$$\frac{\partial I}{\partial t} + \frac{\partial J}{\partial x} = f(I,x,t). \tag{5.4.1}$$

J results from direct social links and many other social factors. For example, in Digg network, besides the fact that a follower votes for news posted by its followee, a user can also vote for any news that he/she is interested in while the news is promoted to the front page, or through search engines provided by the network. In Twitter, the symbol # followed by a few characters, called a hashtag, is used to mark keywords or topics in a tweet. With hashtag symbols anyone can search for the set of tweets that contain one or more specific hashtags. It is estimated that Twitter handles 1.6 billion search queries per day [46]. The use of hashtags increases propagation of tweets. Also Twitter users can send @-messages publicly to a specific user by including the "" character before the receiving person's username in their tweets.

This unstructured phenomenon "jumps" across the network and appears at a seemingly random node [84]. The action results from the relevance of the content of information rather than the structure of the follower graph of a network.

Often, information flows from high density to low density and therefore a simple expression of flux J can be

$$J = -d\frac{\partial I}{\partial x}, \tag{5.4.2}$$

which results from a principle analogous to Fick's law [82] in biology or physics. The minus sign indicates the flow is down the gradient, and d represents the popularity of information that promotes the spread of the information through non-structure based activities such as search. For now d can be viewed as an average and therefore is a constant. In general, it may be dependent on u, x, t. Now we obtain the following PDE model to describe information flow:

$$\frac{\partial I}{\partial t} = d\frac{\partial^2 I}{\partial x^2} + f(I, x, t). \tag{5.4.3}$$

Table 5.1 Interpretations of Eq. (5.4.3) on online social networks with abstract distance and mathematical biology with physical distance

Symbol	Online social networks	Mathematical biology
$J = -d\frac{\partial^2 I}{\partial x^2}$	External influence	Diffusion process (Random walk)
$f(I, x, t)$	Internal influence	Local growth (Birth and death)

Internal influences can be viewed as the growth of population due to local growth in mathematical biology. External influences is similar to the diffusion process in mathematical biology and behaves in a manner of random walk. Extending PDEs into the context of social media, we capture the similarity and difference between spreading of epidemics in biology and information diffusion in online social networks. Table 5.1 compares the difference of the interpretations of PDE models in both mathematical biology and online social networks in the setting of Fig. 5.2.

This framework builds a new architecture for modeling information diffusion in online social networks with partial differential equations. The new architecture engages transdisciplinary approaches involving mathematics, computer science, and social sciences. In particular, clustering algorithms will be used to define abstract cyber-distance for developing partial differential equation models. In the next few chapters we will develop more partial differential equation models based on (5.4.2) and (5.4.3).

5.5 Diffusive Logistic Model

In this section we review the diffusive logistic model introduced in [128] for characterizing the temporal and spatial patterns of information cascading over social media. The model shall be validated with the dataset from Digg. The logistic model is believed to be the simplest nonlinear model to capture population dynamics where the rate of reproduction is proportional to both the existing population and the number of available resources [82]. Innovation diffusion models (2.2.1) describe the diffusion of innovations within a population. In addition, they have been widely used to describe and predict various population dynamics such as bacteria and tumor growth over time [82]. The structure-based process in Fig. 5.3 is modeled with a simple nonlinear logistic equation. Logistic equation is defined as follows. Denoting with N the population at time t, with r the intrinsic growth rate and with K the carrying capacity that gives the upper bound of N, the population dynamics are governed by

$$\frac{dN}{dt} = rN(1 - \frac{N}{K}), \tag{5.5.1}$$

where $\frac{dN}{dt}$ is the first derivative of N with respect to t. In the context of online social networks, the term $rN(1 - \frac{N}{K})$ describes the impact of the network structure on the growth of $I(x, t)$, the density of influenced users at the distance x during time t, and r reflects the decay of news influence with respect to time t.

While some information can take a longer period to spread in social media [19], news diffusion in social media is time-sensitive and the influence of news stories decays drastically as time elapses. Figure 5.4 illustrates the spread of the most popular story in the Digg dataset in the temporal perspective. It shows that interests in news decay exponentially over time. The x-axis is the distance, y-axis is the density of the influenced users. Each line represents the density at time t where t is 1, 2 h, and up to 50 h after the submission of the initial news. The gap of density decreases over time. From our experiments, exponential functions of decay seem plausible for modeling the rapid decay of news with respect to time. The decay process can be modeled by the following ordinary differential equation:

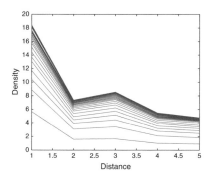

Fig. 5.4 Density of influenced users over 50 h with friendship hops as distance

$$\frac{dr(t)}{dt} = -\alpha r(t) + \beta$$

$$r(1) = \gamma,$$

(5.5.2)

where $\frac{dr(t)}{dt}$ is the rate of change of r with respect to time t, and α is the decay rate, γ is the initial rate of influence. β represents the residual rate as time increases, which can be very small. Solving for r in (5.5.2), we obtain

$$r(t) = \frac{\beta}{\alpha} - e^{-\alpha(t-1)}(\frac{\beta}{\alpha} - \gamma).$$

(5.5.3)

Based on the general spatio-temporal model in Sect. 5.4, we arrive at the following diffusive logistic equation:

$$\frac{\partial I}{\partial t} = d\frac{\partial^2 I}{\partial x^2} + rI(1 - \frac{I}{K})$$

$$I(x, 1) = \phi(x), \ \ l \le x \le L$$

$$\frac{\partial I}{\partial x}(l, t) = \frac{\partial I}{\partial x}(L, t) = 0, \ \ t \ge 1,$$

(5.5.4)

where

- I represents the density of influenced users with a distance of x at time t;
- d represents the popularity of information which promotes the spread of the information through non-structure based activities such as search (external influence);
- r represents the intrinsic growth rate of influenced users with the same distance, and measures how fast the information spreads (internal influence);
- K represents the carrying capacity, which is the maximum possible density of influenced users at a given distance;

- L and l represent the lower and upper bounds of the distances between the source s and other social network users;
- $\phi(x) \geq 0$ is the initial density function, which can be constructed from historical data of information spreading. Each information item has its own unique initial function;
- $\frac{\partial I}{\partial t}$ represents the first derivative of I with respect to time t;
- $\frac{\partial^2 I}{\partial x^2}$ represents the second derivative of I with respect to distance x;

$\frac{\partial I}{\partial x}(l,t) = \frac{\partial I}{\partial x}(L,t) = 0$ is the Neumann boundary condition [82], which means no flux of information across the boundaries at $x = l, L$. This assumption is plausible for social media since the users are clustered in a number of groups U_x. We also assume $\phi(x) \geq 0$ is not identical to zero and the maximum principle implies that (5.5.4) has a unique positive solution $I(x,t)$ and $0 \leq I(x,t) \leq K$.

5.5.1 Initial Density Function Construction

In general, we assume that the initial density function is given and can be constructed using the data collected from the initial stage of information diffusion. Specifically, ϕ is a function of distance x which captures the density of influenced user at distance x at the initial time when a news story is submitted. In online social networks, it is possible to only observe discrete values for the initial density function, because the distance x is discrete. The initial density is the influenced user distribution when time $t = 1$. As in [128], we apply an effective mechanism available in Matlab cubic spline package, called *cubic splines interpolation* [33], to interpolate the initial discrete data in constructing $\phi(x)$. Using this process, a series of unique cubic polynomials are fitted between each of the data points, with the stipulation that the obtained curve is continuous and smooth. Hence $\phi(x)$ constructed by the cubic splines interpolation is a piecewise-defined function and twice continuous differentiable. After cubic splines interpolation, we simply set the two ends to be flat to satisfy the second requirement since in this way the slopes of the density function $\phi(x)$ at the left and right ends are zero.

5.5.2 Accuracy of Diffusive Logistic Model

In this subsection, we evaluate the performance of the proposed linear diffusive model by comparing the density calculated by the model with the actual observations in the Digg dataset. We first present the model accuracy for the most popular news story. Accuracy of the predicated value of a model against an actual value is defined as follows:

$$\text{model\_accuracy} = 1 - \frac{|\text{predicted\_value} - \text{actual\_value}|}{\text{actual\_value}}. \qquad (5.5.5)$$

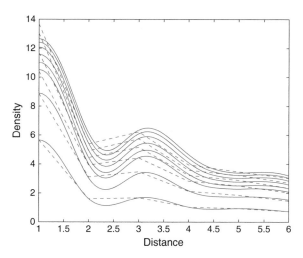

Fig. 5.5 Prediction of (5.5.4) vs. real data of story 1 with 24,099 votes

We numerically solved the model with Matlab. Figure 5.5 illustrates the predicted results for a sample news story (story 1) with the proposed model, where the x-axis is the distance measured by *friendship hops*, while the y-axis represents the density of influenced users within each distance. The solid lines denote the *actual* observations for the density of influenced users for a variety of time periods (i.e., 1, 2, 3, 4, and 5 h), while the dashed lines illustrate the *predicted* density of influenced users by the model. As we can see, the proposed model can accurately predict the density of influenced users with different distance over time. The values of the two parameters K and d in this case are 25 and 0.01. $r(t) = 1.4e^{-1.5(t-1)} + 0.25$. Table 5.2 gives the numerical value of Fig. 5.5. It is clear that the model has high precision in terms of prediction.

5.6 Linear Diffusive Model

The logistic growth model in [128] is the first spatial model to account for the phenomenon in which the initial stage of the increase of influenced users is approximately exponential; then, as saturation begins, the growth slows, and eventually, growth stops. It can achieve a high accuracy as we discuss in the previous section. In this section, we present a more simple linear function to model the growth of influenced users in online social networks [130]. The linear model takes into account the effects of heterogeneity in cyber-distance and news decay with respect to time. As indicated in Fig. 3.3, the distribution of the density of influenced users in distance is not homogeneous. The majority of users are in the groups with distances 3 and 4. This heterogeneity in distance leads to the assumption that the growth function is dependent on location x. The concavity of the shape of Fig. 3.3 further suggests that we can use the following concave down quadratic function $h(x)$ to describe this heterogeneity in distance.

Table 5.2 Prediction accuracy of (5.5.4) with friendship hop as distances for story $s1$

Distance	Average	$t = 2$	$t = 3$	$t = 4$	$t = 5$	$t = 6$
1	98.27%	97.47%	97.74%	97.48%	99.55%	99.09%
2	86.99%	93.59%	96.63%	87.16%	80.80%	76.78%
3	90.28%	83.23 %	87.98%	90.99%	93.35%	95.94%
4	92.98%	86.75%	91.39%	99.00%	95.68%	92.06%
5	93.77%	89.05%	91.61%	97.79%	97.92%	92.49%
6	94.56%	90.03%	89.48%	96.04%	97.57%	99.67%

$$h(x) = -(x - \rho)(x - \sigma). \tag{5.6.1}$$

The coefficient of x^2 in $h(x)$ is scaled to be -1, and $h(x)$ reflects the rate of the change of influenced users with respect to distance x. The simplest way to model the growth of influenced users as linear function of I. Let

$$f = r(t)h(x)I.$$

We can think of $r(t)$ as the average of all distances, and likewise, $h(x)$ as the average of all times. Thus, combining the structure-based process (5.3) and the growth process together gives the following linear diffusive equation:

$$\frac{\partial I}{\partial t} = d\frac{\partial^2 I}{\partial x^2} + r(t)h(x)I$$
$$I(x, 1) = \phi(x), \quad l < x < L \tag{5.6.2}$$
$$\frac{\partial I}{\partial x}(l, t) = \frac{\partial I}{\partial x}(L, t) = 0, \quad t > 1.$$

5.6.1 Accuracy of Linear Model

We evaluate the performance of the proposed linear diffusive model by comparing the density calculated by the model with the actual observations in the Digg dataset. We numerically solved the model with Matlab. Figure 5.6a illustrates the performance of the linear diffusive model for the most popular

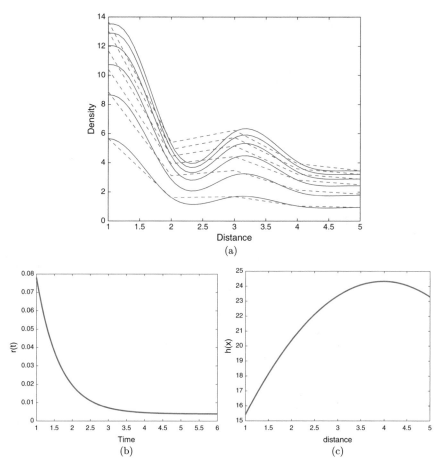

Fig. 5.6 Simulations for the most popular news story in the Digg dataset. Predicted (blue, solid) vs. actual data (red, dotted). (**a**) Predicted (blue, solid) vs. actual data (red, dotted). (**b**) $r(t)$. (**c**) $h(x)$

story, $s1$. Figure 5.6b gives the shape of $r(t)$ and Fig. 5.6c gives the shape of $h(x)$. $h(x)$ is a concave down function with peak between 3 and 4, which is related to the neighbor distribution illustrated in Fig. 3.3. The shape of $h(x)$ suggests that there may exist highly influential users or opinion leaders at distance 4 from the submitter of news $s1$. The corresponding parameters are listed in Fig. 5.7a. The parameters are adjusted manually to best fit the actual data. The diffusion constant d is relatively small because d is the average diffusion rate for all distances. It also suggests that the structure-based process has a dominating effect on the information diffusion process. Further, α, β, γ determine the shape of $r(t)$; and ρ and σ determine the peak of $h(x)$. The average accuracy at different distances is calculated for time $t = 2, \ldots, 6$, provided in Fig. 5.7b. The model can achieve high accuracy across distances.

Parameter	value
d	0.0020
α	1.5526
β	0.0059
γ	0.0780
ρ	-0.9478
σ	8.9149

(a)

Distance	Average
1	97.88%
2	97.27%
3	97.44%
4	96.20%
5	98.25%
Overall	97.41%

(b)

Fig. 5.7 Accuracy of (5.6.2) for the most popular news story in the Digg dataset. (a) Parameter values. (b) Model accuracy

We also study the accuracy of the model for describing all news stories in the Digg dataset and examine whether the model can capture the heterogeneity features in information diffusion over the Digg network [130]. We explore the overall accuracy of the linear diffusive model for all 133 news stories in the Digg dataset with more than 3000 votes. Our results in [130] illustrate that about 13% of news stories can be described with accuracy higher than 90%. In total, about 60% of news stories can be described with accuracy higher than 80%. The simulation is performed with a Matlab auto fitting program. If we manually adjust parameters for each individual news story, higher accuracy can be achieved. For example, for the most popular news story, with manually adjusted parameters, the average accuracy can reach 97.41%, while with the automated parameter selection, the average accuracy is greater than 90%. The high accuracy across all news stories with more than 3000 votes shows strong evidence that the linear diffusive model captures the heterogeneity diffusion patterns of news and can be used as an effective approach to describe the news spreading in Digg.

5.7 Logistic Model with Biased Diffusion

In the previous two sections, we assume that the diffusion coefficient d in the flux formula

$$J = -d\frac{\partial I}{\partial x}$$

is a constant. In fact, because of the spatial heterogeneity of online users in social media, d may be dependent on the distance x from the source. In general, d may be a decreasing function of x since interactions between different groups U_x decrease dramatically as x increases. Therefore, we use an exponential function

$$d = de^{-bx}$$

to model the effect of spatial heterogeneity of online users in the external influence in Fig. 5.3. Thus we arrive at the following equation:

$$\frac{\partial I}{\partial t} = \frac{\partial (de^{-bx}I_x)}{\partial x} + r(t)I(h(x) - \frac{I}{K})$$

$$I(x,1) = \phi(x), \quad l < x < L$$

$$\frac{\partial I}{\partial x}(l,t) = \frac{\partial I}{\partial x}(L,t) = 0, \quad t > 1 \tag{5.7.1}$$

$$r(t) = A + Be^{-Ct},$$

where

- d represents the popularity of information;
- b represents the decay of the popularity of information with respect to the friendship structure in social networks;
- K represents the carrying capacity, which is the maximum possible density of influenced users at a given distance;
- $h(x)$ represents the heterogeneity of growth rate in distance x.

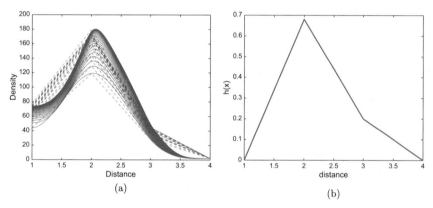

Fig. 5.8 Accuracy of (5.7.1) for a photo tweet from President Obama at Twitter. (a) Predicted (blue, solid) vs. actual data (red, dotted). (b) $h(x)$

Distance	Average
1	98.39%
2	98.75%
3	94.11%
4	99.31%
Overall	97.64%

Fig. 5.9 Accuracy of (5.7.1) at $x = 1,2,3,4$ for the photo tweet from President Obama

5.7.1 Accuracy of Logistic Model with Biased Diffusion

We evaluate the performance of the logistic model with variable diffusion by comparing the density calculated by the model with the actual observations in the Twitter dataset. We choose a photo tweet from President Barack Obama on December 13, 2012 as an example to verify accuracy of the model. Based on the Twitter user list and social network graphs collected in October 2009, we find 383 Twitter users have retweeted this photo. Among these 383 users, 96 have a distance of 1 to the source of the tweet message, and 226, 58, and 2 users have a distance of 2, 3, and 4, respectively. We compare the number of users retweeting the photo at the different distances and a time frame from $t = 1$ to $t = 15$.

We numerically solved the model with Matlab. Figure 5.8a illustrates the predicting results for the photo tweeted by President Barack Obama with the proposed model for the time frame from $t = 1$ to $t = 15$ h, where the x-axis is the distance measured by *friendship hops*, while the y-axis represents the number of retweeted users within each distance. The red lines represent the actual number of people retweeting the photo at various time increments. The blue curves represent the model used to predict the information diffusion based on the PDE model. The red-dotted lines denote the *actual* observations for the number of retweeted users. The mathematical model representing the diffusion of President Obama's tweet reaches an overall accuracy of 97.64%, shown in Fig. 5.9. These results were obtained with $d = 1, b = 3, K = 300$, $r(t) = 0.3 + e^{-2t}$, $h(x)$ function shown in Fig. 5.8b. Figure 5.8b illustrates a peak at $x = 2$, which can indicate that President Obama's tweeted photo is most popular within the Twitter users at distance 2.

Therefore, the spatial models (5.5.4), (5.6.2), and (5.7.1) can achieve high accuracy. While (5.6.2) is a linear model and captures the behavior of news spread within a few hours, nonlinear models (5.5.4) and (5.7.1) can predict news spread for a longer time frame. $h(x)$ in (5.6.2) and (5.7.1) reflects the spatial heterogeneity of online users with respect to distance x. In (5.5.4) it is assumed that $h(x)$ is constant. Equation (5.7.1) takes into account the fact that the diffusion coefficient d is a decreasing function of x, which is a constant in (5.5.4). From the experiments above, all three models can achieve high accuracy.

Chapter 6
Modeling Complex Interactions

Abstract In this chapter we present a number of partial differential equation models to describe complex interactions among users in online social networks. We discuss an information diffusion model initiated from several sources and extend spatial cooperation systems for population dynamics to information diffusion where two or more pieces of information promote each other. We adopt the spatial Lotka–Volterra competition model to information diffusion and discuss its application in modeling the market share competition of smartphone operating systems. Finally, we extend epidemiological models to information diffusion where spatial arrangement is present.

6.1 Introduction

The information diffusion process over online social networks may be influenced by complex interactions between users and information, e.g., information originating from more than one source and competing information from different political campaigns. Myers et al. [83] (also see Guille et al. [38]) study cooperation and competition in information diffusion in the Twitter network using a statistical model. In this chapter, we present a number of simple models to illustrate how to incorporate complex interactions in the spatial-temporal setting. Some of the results presented here are based on the authors' papers [96, 148].

© Springer Nature Switzerland AG 2020   59
H. Wang et al., *Modeling Information Diffusion in Online Social Networks
with Partial Differential Equations*, Surveys and Tutorials in the Applied
Mathematical Sciences 7, https://doi.org/10.1007/978-3-030-38852-2_6

6.2 Information Diffusion Initiated from Multiple Sources

6.2.1 Distance Metric

In an online social network, the same information could originate from multiple sources. Often, breaking news stories, emergency events, and controversial topics are initiated by a number of different news sources, e.g., various users tweeting the final result for a sport game in Twitter. Reporting of news from diverse sources often increases its spreading speed and coverage. It is more practical and important to understand the diffusion patterns of multi-source information in social media.

In a paper [96] the authors use the linear diffusion model (5.6.2) to predict the information diffusion process of multi-source news spread in Digg. This study examines the basic characteristics of the diffusion process of multi-source information. Unlike single-source information, multiple-source information originates from a set of initiators in OSNs (online social networks). For example, two Digg users first submit a certain news story at the same time before the rest of other users in OSNs, and then the same information cascades in parallel along both temporal and spatial dimensions. Thus new models are required to investigate the diffusion process of information that is initiated simultaneously by two or more sources, referred to as *an information diffusion from multiple-sources problem*: given an information item initiated from a set of multiple sources $S = s_1, s_2, \ldots \ldots, s_m$, what is the density of influenced user, $d(x, t)$, at the distance of x from the multi-sources after a period of time t?

The distance metric in [96] is intuitively defined as the minimum shortest path between a user and multiple sources. The distance definition reflects the fact that while a user of social media could be influenced by a set of news sources through different paths, the nearest source to which the user has the minimum friendship hop has the highest probability of influencing the given user's behavior, because of the smallest number of friendship hops. The prediction accuracy of the linear diffusion model (5.6.2) for multiple news sources is validated with the same dataset from Digg. The experiments in [96] show that the model can describe the most representative news stories initiated from multiple sources with an accuracy higher than 90%, and can achieve an average accuracy around 75% across all multi-source news stories in the dataset. These results confirm that our approach with reaction-diffusion equation is able to describe and predict the spreading patterns of multi-source information with high accuracy.

6.2.2 Experiment Results

In [96] we identify a collection of 1433 news stories that are simultaneously initiated from two or more sources. Based on the number of the sources each story has, we classify these 1433 news stories into six groups: 2-source, 3-source, 4-source, 5-source, 6-source, and 8-source. The six groups include 1045, 304, 64, 16, 3, and 1 news stories, respectively. Now we validate the accuracy of the *linear diffusive model*

$$\frac{\partial I}{\partial t} = d\frac{\partial^2 I}{\partial x^2} + r(t)h(x)I$$
$$I(x,1) = \phi(x), \quad l < x < L \qquad (6.2.1)$$
$$\frac{\partial I}{\partial x}(l,t) = \frac{\partial I}{\partial x}(L,t) = 0, \quad t > 1$$

by comparing the densities calculated by the model with the actual values observed in the Digg dataset. We quantitatively measure the predicting accuracy of the model, $f_{accuracy}$ as follows:

$$f_{accuracy} = 1 - \frac{|v_p - v_a|}{v_a}, \qquad (6.2.2)$$

where v_p denotes the density predicted by the model while v_a denotes the actual value from the real Digg dataset. Clearly, $0 \leq f_{accuracy} \leq 1$.

Figure 6.1a–f illustrate the accuracy of the model for the most popular news stories s1 to s6 in each group of multiple-source news stories. As shown in these figures, the model achieves a high accuracy in predicting the density of influenced users over time across six different news stories initiated from different numbers of sources.

To evaluate the accuracy of the model on all other news stories, we run the model on all 1433 news stories in our dataset. Table 6.1 illustrates the results on the accuracy of the model on all these news stories as well as on the most popular news for each group. The first column denotes the group of multiple-source news stories; the second and third columns show the most popular news story for each group and the prediction accuracy. The last two columns summarize the total number of news stories in each group and the average prediction accuracy among all the news stories in the same group. Apparently, the model can achieve over 90% accuracy for the top news from all groups. More importantly, this model achieves very high prediction accuracy for other news stories as well. The average accuracy of all 1433 news stories is 76.25%, and the average accuracies for all groups are higher than 70%. These findings confirm the prediction capability of the linear diffusion model on news stores initiated from single sources as well as news stories initiated from multiple sources.

The PDE-based diffusion models address the spatial-temporal problem of information diffusion. Specifically, the models successful describe and predict

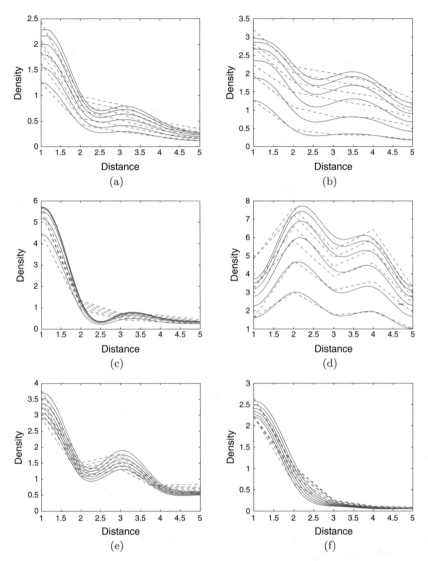

Fig. 6.1 The prediction accuracy of a linear diffusion model on six popular news initiated from multiple sources. The dashed line denotes the actual observation for the density of influenced users over time, while the solid one denotes the density calculated by the model. (**a**) News story s1. (**b**) News story s2. (**c**) News story s3. (**d**) News story s4. (**e**) News story s5. (**f**) News story s6

the density of influenced users, $d(x, t)$, at a distance of x from the information source after a certain period of time t. The experimental results based on Digg news stories demonstrate the capability and high accuracy of the models in describing the process of information diffusion initiated from a single source or from multiple sources.

Table 6.1 Prediction accuracy for all news stories and the most popular news stories for each group

Group	Story	Accuracy	Total news stories	Average accuracy
2-Source	s1	93.43%	1045	76.55%
3-Source	s2	93.69%	304	75.70%
4-Source	s3	92.61%	64	73.63%
5-Source	s4	90.97%	16	77.82%
6-Source	s5	94.49%	3	76.41%
8-Source	s6	95.55%	1	95.55%

6.3 Cooperating Diffusion Process

6.3.1 Cooperative Systems

The effect of multiple sources can be also viewed as mutualism in biological systems where two or more organisms of different species biologically interact in a relationship in which each individual derives a fitness benefit (i.e., increased or improved reproductive output). Such interactions in biological systems are known as cooperation systems [82]. We attempt to use dynamical equations to predict information diffusion in social media with multiple sources and channels. If we consider two news sources, then the two pieces of news spread with logistic growth independently along with additional effect from another. Let u_i be the density of online users who have been influenced by information from different source $s_i, i = 1, 2$. The positive effects can be modeled by terms $\alpha_1 u_1 u_2$ and $\alpha_2 u_1 u_2$. As a result, the following simple model can be a starting point to address the impact between multiple sources:

$$\frac{\partial u_1}{\partial t} = \frac{\partial (d_1 e^{-b_1 x} (u_1)_x)}{\partial x} + r_1(t) u_1 (h_1(x) - \frac{u_1}{k_1}) + \alpha_1 u_1 u_2$$
$$\frac{\partial u_2}{\partial t} = \frac{\partial (d_2 e^{-b_2 x} (u_2)_x)}{\partial x} + r_2(t) u_2 (h_2(x) - \frac{u_2}{k_2}) + \alpha_2 u_1 u_2,$$

(6.3.1)

where

- d_1, d_2 represent the popularity of the two pieces of information;
- b_1, b_2 represent local decay rates of the popularity of the two pieces of information along the distance;
- $r_i(t), i = 1, 2$ represents the intrinsic growth rate of influenced users within the same distance and measures how fast the information spreads within the same cluster;
- $k_i, i = 1, 2$ represents the carrying capacity, which is the maximum possible density of influenced users;
- α_1 measures the positive effect of news u_2 on u_1 and α_2 measures the positive effect of news u_1 on u_2;
- $h_i(x)$ represents the heterogeneity of growth rate in distance x.

6.3.2 Modeling the Interaction of Mobile Phones and Mobile Applications

We will apply system (6.3.1) to describe how often the two key words "Apps" (Mobile Applications) and "Mobile Phones" are searched in Google. The real dataset comes from Google Trends. Google Trends is a public web facility based on Google search. It provides how often a particular search-term is entered relative to the total search-volume across various regions of the world, and in various languages. Google Trends also allows the user to compare the volume of searches between two or more terms. An additional feature of Google Trends is its ability to show news related to the search-term overlaid on the chart, showing how new events affect search popularity. Many groups use Google Trends to analyze data for market research, keyword research, stock market, and traffic information.

Here we compare "Apps" (Mobile Applications) and "Mobile Phone" to see how the two words promote each other. Many American adults report that they get at least some local news and information on their cellphones or tablet computers. Many news organizations are relying on mobile platforms to provide new ways to generate revenue in local markets. We want to look at how information spreads from a state to another state. We are interested in how far the information spreads in a certain period of time based on distance from a given state. Using Google Trends we select from March 6, 2011 to July 17, 2011 to compare "Apps" (Mobile Apps) and "Mobile Phones." We organize this data from the West coast to the East coast with an assumption that the related information travels from the West coast to the East coast.

Figure 6.2 shows the predicting results for (6.3.1), with x-axis representing the distance from California. $1, 2, 3, 4, 5, 6, 7, 8, 9$ represent California, Texas, Louisiana, Illinois, Mississippi, Tennessee, Georgia, Florida, and North Carolina, respectively. The solid lines denote the *actual* observations for the index of the two key words for a variety of time periods (i.e., 1 week, 2 week, 3 week,

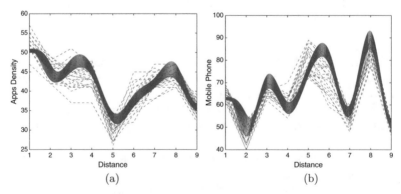

Fig. 6.2 Predicted (blue, solid) for (6.3.1) vs. actual data (red, dotted). (**a**) Apps. (**b**) Mobile phone

4 week, etc.), while the dashed lines illustrate the *predicted* value from (6.3.1). In general, the proposed model can accurately predict the index with different distances over time. The values used for the parameters d_1, d_2, b_1, b_2, k_1, k_2, and α_1, α_2 in this case are $1, 1$, $3.08028, 3.07628$, $2000.34192, 1999.14992$, $0.00005, 0.00010$, $r_1(t) = 0.0001e^{-19t}$, $r_2(t) = 0.0001e^{-20t}$, $h_1(x) = 0.00001$, and $h_2(x) = 0.00050$, respectively. Accuracy for u_1 and u_2 are 94.34 and 92.09%, respectively. The experiments with other time frame lead to similar observations.

In summary, the experiment results with Google Trends datasets show that our proposed PDE-based model effectively predicts the competing process of information cascading on social media.

6.4 Competing Diffusion Process

6.4.1 Competition Systems

Different political views or product campaigns compete with each other to maximize its own influences on people through online social networks. To characterize the density of users that are influenced by a certain competition, we may study competing news by using the Lotka–Volterra competition models [82]. Competition models have been extensively studied in mathematical biology [82] where interactions between organisms or species can reduce the presence of one or the other due to limited resources (such as food, water, and territory).

A simple mathematical model to describe news competition may be the same as Eq. (6.3.1) except that the effect of the competition is negatively proportional to the number of influenced users.

$$
\begin{aligned}
\frac{\partial u_1}{\partial t} &= \frac{\partial(d_1 e^{-b_1 x}(u_1)_x)}{\partial x} + r_1(t)u_1(h_1(x) - \frac{u_1}{k_1}) - \alpha_1 u_1 u_2 \\
\frac{\partial u_2}{\partial t} &= \frac{\partial(d_2 e^{-b_2 x}(u_2)_x)}{\partial x} + r_2(t)u_2(h_2(x) - \frac{u_2}{k_2}) - \alpha_2 u_1 u_2,
\end{aligned}
\tag{6.4.1}
$$

where

- d_1, d_2 represent the popularity of the two pieces of information;
- b_1, b_2 represent local decay rates of the popularity of the two pieces of information along the distance;
- $r_i(t), i = 1, 2$ represents the intrinsic growth rate of influenced users within the same distance and measures how fast the information spreads within the same cluster;
- $k_i, i = 1, 2$ represents the carrying capacity, which is the maximum possible density of influenced users;

- α_1 measures the negative effect of u_2 on u_1 and α_2 measures the negative effect of u_1 on u_2;
- $h_i(x)$ represents the heterogeneity of growth rate in distance x.

The model captures the logistic growth of each information and also quantifies the effect of the competition on the diffusion process. Some relevant and practical questions, such as which news can win the competition, can be answered by studying the competition models in Sect. 7.5.5.

6.4.2 Modeling the Ecosystem of Smartphone Operating Systems

We shall use a spatial Lotka–Volterra competition model to study and predict market share of mobile phone operating systems. Currently there are two major operating systems: Android (Google) (u_1), iPhone (Apple) (u_2). Using Google Trends we choose the two topics "iPhone" and "Android" to compare how often the two words "iPhone" and "Android" are searched in Google. We are interested in the spread of related information over a period of time among users aggregated by state. We organize the Google Trends data from the West coast to the East coast because the main headquarters for Apple and Google are located on the West coast. Hence the news about Android and iPhone should start from the West coast.

The construction of initial conditions follows the method outlined in Sect. 5.5.1. The x-axis values represent the distance from California. $1, 2, 3,$ $4, 5, 6, 7, 8, 9$ represent California, Texas, Louisiana, Illinois, Mississippi, Tennessee, Georgia, Florida, and North Carolina, respectively. The values used for the parameters $d_1, d_2, b_1, b_2, k_1, k_2,$ and α_1, α_2 in this case are $1, 1, 3.01776,$ $3.00109,$ $2014.01720, 2008.14445,$ $0.00090, 0.00050,$ $r_1(t) = 0.10182e^{-1.99377t},$ $r_2(t) = 0.09974e^{-2.02209t},$ $h_1(x) = 0.22,$ and $h_2(x) = 0.11,$ respectively. The time frame is from May 2011 to July 2011. With these parameters and data, the simulation accuracy is above 90%.

We choose the time frame from May 2011 to July 2011 because Apple released a number of major products, applications, and operating systems in this time frame. Steve Jobs, Apple's CEO claimed that iCloud can keep important information and content up to date across all of one's devices. In the meantime, Google introduced new phone bigger, faster, thinner, lighter. However, one of the greatest technological advances in 2011 was Apple's iCloud. This also explains why the search index for iPhone is higher than Android during this period.

6.5 Spatial Epidemiological Models

It is commonly accepted that dynamical models to describe disease spreading can be used to study information. There is a considerable body of research work modeling information diffusion based on techniques for modeling infectious diseases. In general, epidemiological models are neither cooperative nor competitive and they are more challenging to study. Most of the applications of epidemiological models on social media have largely concentrated on collective analysis and deal with only ordinary differential equations. With the new metric concept between users we introduced in previous sections, spatial effects can be incorporated and partial differential equations can play an important role in studying information diffusion. In particular, spatial models take into consideration the external influence. Many similar models in epidemiology can be further modified and expanded to study information diffusion in online social networks. For social media, S represents the density of susceptible users at time t and distance x in U_x and I represents the density of influenced users at time t and distance x in U_x. The following SI model is a simple example of how spatial infectious disease modeling can be used to study online social networks.

As we explain in Sect. 2.3, for a deterministic epidemiological model, the nodes of a network are classified into two classes (SI) and focus on the evolution of the proportions of nodes in each class, where S stands for "susceptible," I for "infected" (i.e., adopted the information). Nodes in the S class switch to the I class due to influence of their neighbor nodes. The percentage of nodes in each class is expressed by simple differential equations. The model assumes that every node has the same probability to be connected to another and thus connections inside the population are made at random. Now we arrive at the following spatial SI model:

$$\frac{\partial S}{\partial t} = d_1 \frac{\partial^2 S}{\partial x^2} - r(t) \frac{SI}{S+I}$$
$$\frac{\partial I}{\partial t} = d_2 \frac{\partial^2 I}{\partial x^2} + r(t) \frac{SI}{S+I},$$

$$(6.5.1)$$

where $r(t)$ is the rate of influence. $S+I$ appears in the denominator because it is not necessarily constant in spatial models. One important concept to describe the interactions between user groups is the rate of influence $r(t)$ which is similar to the force of infection in epidemiology. The choice of the rate of influence is largely dependent on news and user classifications. Data mining techniques and graphical model can significantly improve the selection of the parameters.

Zhu et al. [148] introduced a novel epidemic-like model with both discrete and nonlocal delays for investigating the spatial-temporal dynamics of rumor propagation in social networks. Applying the linear approximation method of nonlinear systems, we establish sufficient conditions for the existence of Hopf bifurcation at its equilibrium points and identify some sensitive parameters in the process of rumor propagation.

6.6 Multiple Communication Channels

Most of the research on information diffusion is limited to studying information diffusion from isolated single media sites or factors. However, many online social networks involve multiple connections between any pair of nodes (such as friendship, family relationship, geographic location, and other demographic characteristics). In addition, news propagation in one social media site often affects that of another social media site, for example, Facebook vs. Twitter. A theoretical framework for understanding of information flow in online social networks over multiple communication channels is still missing. With the framework to model information diffusion in social media with PDE models, it is possible to use partial differential equations defined on multidimensional domains to analyze, characterize, and predict spatial-temporal dynamics of information diffusion with multiple communication channels. The governing partial differential differentials can be systems of reaction-diffusion equations defined on multidimensional domains. For example, with almost the same setting as in (5.5.4), a simple diffusive logistic equation defined on multidimensional domain Ω may be able to describe information diffusion over social media with multiple communication channels

$$
\begin{aligned}
\frac{\partial I}{\partial t} &= d\Delta I + rI(1 - \frac{I}{K}) \\
I(x, 1) &= \phi(x), \quad x \in \Omega \\
\frac{\partial I}{\partial n} &= 0, \quad \text{on } \partial\Omega \times (1, \infty),
\end{aligned}
\tag{6.6.1}
$$

here Δ is the Laplacian operator defined on a multidimensional domain $\Omega \subset R^n$, and n is the number of independent communication channels we are interested in. For example, if we are interested in studying information diffusion with two unrelated channels, a region in R^2 may be plausible to describe the combined effect of the two channels on the spread of news in social media. Systems of equations such as (6.3.1) or (6.5.1) can also be defined on multidimensional domains to model information diffusion with multiple communication channels.

Chapter 7
Mathematical Analysis

Abstract In this chapter we discuss a free boundary problem for reaction-diffusion logistic equations in online social networks. Specifically, we discuss several bifurcation and stability results for a nonautonomous diffusive logistic model in online social networks with Robin boundary conditions. In addition, we present Hopf bifurcation and spatial patterns of an epidemic-like rumor model for online social networks. Finally, we present traveling wave solutions of diffusive models and give long-term propagation rates of information diffusion in online social networks.

7.1 Introduction

We have presented a number of partial differential equation models to characterize spatial-temporal patterns in social media. These spatial-temporal models are reaction-diffusion equation models built on intuitive cyber-distance in online social networks. The basic mathematical properties of these models such as the existence, uniqueness, and positivity of their solution can be established from the standard theorems for parabolic PDEs [17, 114]. More references can be found in [30, 31, 44, 76, 94, 108, 147]. Many other mathematical properties remain to be investigated. Mathematical analysis of the models can not only validate the models, but also reveal insights on information flow over social media. These problems demonstrate fascinating connections between advanced mathematics and online social networks. The extension of applications of PDEs into online social networks presents new opportunities and challenges for mathematicians.

Part of the materials in this chapter is based on the authors' papers including [67] on a free boundary problem, [23] on bifurcation, stability of the

© Springer Nature Switzerland AG 2020   69
H. Wang et al., *Modeling Information Diffusion in Online Social Networks with Partial Differential Equations*, Surveys and Tutorials in the Applied Mathematical Sciences 7, https://doi.org/10.1007/978-3-030-38852-2_7

diffusive logistic model with heterogeneity in distance, [149] on stability and
spatial patterns of an epidemic-like rumor propagation model, and [125, 127]
on traveling wave solutions.

7.2 Free Boundary Problems in Online Social Networks

7.2.1 Free Boundary Problems

In our previous models (5.5.4),(5.6.2), (5.7.1), no flux boundary condition is
assumed with the understanding that no formation flows across the boundaries at $x = l, L$. In general, information may flow across the boundaries. To
describe the change of boundary with respect to time t, [67] proposes and
studies a free boundary model to describe the spreading of news in online
social networks.

We investigate the following diffusive logistic model with a free boundary
for online social networks. The system is given in the following form:

$$\begin{cases} u_t - du_{xx} = r(t)u(1 - \frac{u}{K}), & t > 0, \ 0 < x < h(t), \\ u_x(t,0) = 0, \ u(t, h(t)) = 0, & t > 0, \\ h'(t) = -\mu u_x(t, h(t)), & t > 0, \\ h(0) = h_0, \ u(0, x) = u_0(x), & 0 \le x \le h_0, \end{cases} \tag{7.2.1}$$

where the initial function $u_0(x)$ belongs to $\Sigma(h_0)$ for some h_0 satisfying

$$\Sigma(h_0) = \{\varphi \in C^2([0, h_0]) : \ \varphi'(0) = \varphi(h_0) = 0, \ \varphi(x) > 0 \text{ in } (0, h_0)\}.$$

Here, $u(t, x)$ represents the density of influenced users with distance x at
time t; $x = h(t)$ is the moving boundary to be determined and represents
the spreading front of news (such as a movie recommendation) among users.
$u_x(t, 0) = 0$ means no news traveling in the left side of the region. Hence we
only need to consider the diffusion in the right side of the region. K is the
carrying capacity, d is the diffusion rate, and $r(t)$ is intrinsic growth rate.
Furthermore, we always assume the following condition:

- (A) $r(t)$ is a decreasing function of time t with a positive lower bound,
  i.e., $0 < r_\infty \le r(t) \le r(0)$.

In general, we call $h'(t) = -\mu u_x(t, h(t))$ the Stefan condition, where μ
represents the diffusion ability of the information in a new region. As we
know, the Stefan condition has been used in many areas. For example, it was
used to model the wound healing [20], the melting of ice [108], and the spread
of species [27, 71]. Here we study a free boundary problem for describing the
spread of news in online social networks. In Sect. 7.2.2, we extend (7.2.1) to
a system with multiple information and present numerical results. Finally we
discuss theoretical results in Sect. 7.2.3.

7.2.2 Free Boundary Problems with Multiple Information

In this section, we investigate a multiple information diffusion process with intervention in online social networks. For example, a health intervention from public health agencies can use online social networks to promote behavior that improves mental and physical health, or discourages or reframes those with health risks. A better understanding of multiple information diffusion process with intervention is important. First, information diffusion originated from diverse sources has not been fully understood. Recently, Peng et al. [96] studied information diffusion initiated from multiple sources in online social networks by numerical simulation. There remain many challenging problems in modeling and analysis of more than one information diffusion processes. In addition, increasing numbers of rumors appear in online social networks and our mathematical results may shed light on effectively controlling their spread in online social networks.

Zhu et al. [101] studied a simple case where there are three pieces of information, A, B, and C, sent from different sources to compete for influence on online users. We view information C as intervention from the media or government to control the spread of information A and B. For simplicity, we assume that A and B have no influence on C. Information A and B compete for influence on each other. Figure 7.1 illustrates the interactions among information A, B, and C. As we discuss in Sect. 7.2.1, the free boundary problem for this competition system with intervention can be modeled by the following system:

$$
\begin{cases}
u_t - d_1 u_{xx} = r_1 u(1 - a_1 u - b_1 v - c_1 w), & 0 < x < h(t), t > 0, \\
v_t - d_2 v_{xx} = r_2 v(1 - a_2 u - b_2 v - c_2 w), & 0 < x < h(t)), t > 0, \\
w_t - d_3 w_{xx} = r_3 w(1 - c_3 w), & 0 < x < h(t)), t > 0, \\
u_x(0,t) = v_x(0,t) = w_x(0,t) = 0, & t > 0, \\
u(h(t),t) = v(h(t),t) = w(h(t),t) = 0, & t > 0, \\
h'(t) = -\mu[u_x(h(t),t) + \rho_1 v_x(h(t),t) + \rho_2 w_x(h(t),t)], & t > 0, \\
u(x,0) = u_0(x), v(x,0) = v_0(x), & 0 < x < h_0 \\
w(x,0) = w_0(x), h(0) = h_0, & 0 < x < h_0,
\end{cases}
$$

$$(7.2.2)$$

where μ, ρ_1, ρ_2, and h_0 are given positive constants; $x = h(t)$ is the free boundary, and the initial function $u_0(x), v_0(x), w_0(x) \in \Sigma(h_0)$ for some $h_0 > 0$, where

$$\Sigma(h_0) = \{\phi \in C^2([0,h_0]) : \phi'(0) = \phi(h_0) = 0, \ \phi(x) > 0 \ \text{in} \ [0,h_0)\}. \quad (7.2.3)$$

Here $u(x,t), v(x,t), w(x,t)$ represent the density of influenced users at time t and distance x. $1/a_1, 1/b_2, 1/c_3$ are the carrying capacities, $d_i(i = 1, 2, 3)$ are the diffusion rates, $r_i(i = 1, 2, 3)$ are the intrinsic growth rates, b_1, a_2, c_1, c_2 are the competition or intervention rates. All parameters are positive.

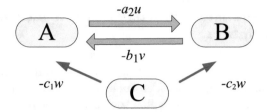

Fig. 7.1 Multiple information diffusion process

Here we only present numerical results on information spreading in online social networks following the approach in [101]. Theoretical results are similar to those in [67] and can be found in [101]. For problem (7.2.2) let

$$d_1 = d_2 = 1, d_3 = 2, r_1 = 1, r_2 = 0.5, r_2 = 2, \rho_1 = 1, \rho_2 = 0.5,$$

$$u_0(x) = v_0(x) = \cos(\frac{\pi x}{2h_0}), w_0(x) = h_0^2 - x^2, h_* = \frac{\pi}{2}\min\{\sqrt{\frac{d_i}{r_i}}, i = 1, 2, 3\} = \frac{\pi}{2}.$$

Now we give two examples on the long-time behavior of solutions (u, v, w) of problem (7.2.2).

Example 7.1. Fix $h_0 = 1(< h_* = \pi/2), a_1 = 1, a_2 = 0.3, b_1 = 0.4, b_2 = 2, c_1 = 0.3, c_2 = 0.4, c_3 = 1$. The numerical solutions of problem (7.2.2) with $\mu = 0.1$ and $\mu = 1$ are shown in Figs. 7.2 and 7.3, respectively. From Fig. 7.2, we

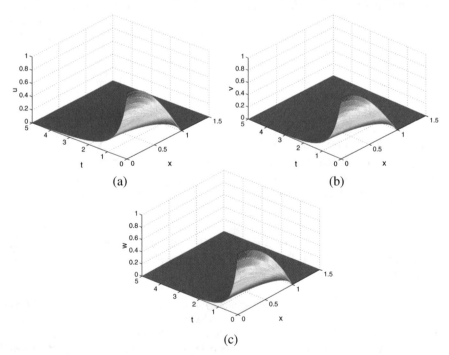

Fig. 7.2 (**a**) u decays to zero quickly. (**b**) v decays to zero quickly. (**c**) w decays to zero quickly

observe that the free boundary $x = h(t)$ increases slowly, and the solutions decay to zero quickly, implying that all information vanish with small expanding capability ($\mu = 0.1$). In Fig. 7.3 the free boundary $x = h(t)$ increases faster than in Fig. 7.2 and the solutions stabilize to positive equilibria, that is, all information spread with large expanding capability ($\mu = 1$).

Example 7.2. Let $h_0 = 1, \mu = 1, a_1 = b_2 = 1, b_1 = a_2 = 0.5, c_1 = 1.5, c_2 = 2, c_3 = 1$. From Fig. 7.4, we find that in (a) and (b) the solutions decay to zero quickly; in (c) the free boundary increases quickly and the solution stabilizes to a positive equilibrium. Thus we conclude that information A, B vanish and C spreads.

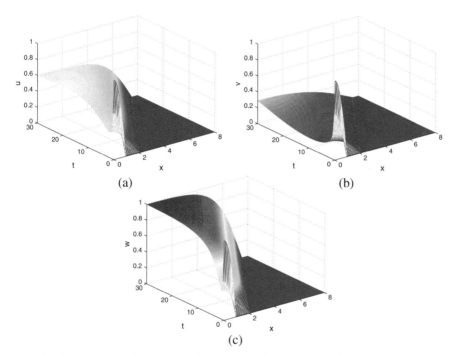

Fig. 7.3 (**a**) u stabilizes to a positive equilibrium. (**b**) v stabilizes to a positive equilibrium. (**c**) w stabilizes to a positive equilibrium

7.2.3 Theoretical Results on Free Boundary Problems

In this section we present a number of theoretical results for (7.2.1). We first present the local existence and uniqueness of the solution of (7.2.1) which can be proved by the contraction mapping theorem as in [27].

Theorem 7.2.1 Let $D = \{(x,t) \in \mathbb{R}^2 : x \in [0, h(t)], t \in [0, \infty)\}$. For any $\alpha \in (0,1)$, problem (7.2.1) admits a unique solution

$$(u, h) \in C^{(1+\alpha)/2, 1+\alpha}(D) \times C^{1+\alpha/2}([0, \infty)).$$

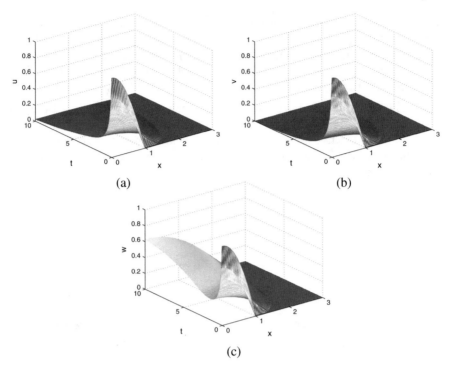

Fig. 7.4 (**a**) u decays to zero quickly. (**b**) v decays to zero quickly. (**c**) w stabilizes to a positive equilibrium

Furthermore,

$$0 < u(t,x) \leq M, \ 0 < h'(t) \leq C$$

for all $t > 0, 0 < x < h(t)$, where C and M are positive constants.

The following comparison principle plays an essential role in this section for estimating solutions of (7.2.1).

Lemma 7.2.2 (Comparison Principle) (*i*) Suppose that $\overline{u} \in C(\overline{D_{T_0}^*}) \cap C^{1,2}(D_{T_0}^*)$, $\overline{h} \in C^1([T_0, T])$ with $D_{T_0}^* = \{(t,x) \in R^2 : 0 \leq T_0 < t \leq T, 0 < x < \overline{h}(t)\}$, and

$$
\begin{cases}
\overline{u}_t - d\overline{u}_{xx} \geq r(t)\overline{u}(1 - \frac{\overline{u}}{K}), & t > T_0, \ 0 < x < \overline{h}(t), \\
\overline{u} = 0, \quad \overline{h}'(t) \geq -\mu\overline{u}_x, & t > T_0, \ x = \overline{h}(t), \\
\overline{u}_x(t,0) \leq 0, & t > T_0, \\
\overline{h}(T_0) \geq h(T_0), \ \overline{u}(T_0,x) \geq u(T_0,x), & 0 \leq x \leq h(T_0).
\end{cases}
\tag{7.2.4}
$$

Then the solution (u,h) of the free boundary problem (7.2.1) satisfies

$$h(t) \leq \overline{h}(t), \ u(t,x) \leq \overline{u}(t,x)$$

for $(t,x) \in [T_0, T] \times [0, h(t)]$.

(ii) Assuming that $\underline{u} \in C(\overline{D^*_{T_0}}) \cap C^{1,2}(D^*_{T_0})$, $\underline{h}(t) \in C^1([T_0, T])$ with $D^*_{T_0} = \{(t, x) \in R^2 : 0 \leq T_0 < t \leq T, 0 < x < \underline{h}(t)\}$, and

$$
\begin{cases}
\underline{u}_t - d\underline{u}_{xx} \leq r(t)\underline{u}(1 - \frac{u}{K}), & t > T_0, \ 0 < x < \underline{h}(t), \\
\underline{u} = 0, \quad \underline{h}'(t) \leq -\mu\underline{u}_x, & t > T_0, \ x = \underline{h}(t), \\
\underline{u}_x(t, 0) \geq 0, & t > T_0, \\
\underline{h}(T_0) \leq h(T_0), \ \underline{u}(T_0, x) \leq u(T_0, x), & 0 \leq x \leq \underline{h}(T_0).
\end{cases}
\tag{7.2.5}
$$

Then the solution (u, h) of the free boundary problem (7.2.1) satisfies

$$
h(t) \geq \underline{h}(t), \quad u(t, x) \geq \underline{u}(t, x)
$$

for $t \in [T_0, T]$, $x \in [0, \underline{h}(t)]$.

Lemma 7.2.2 implies that for any solution u of (7.2.1), we have

Corollary 7.2.3 $u \leq \overline{u}$, where

$$
\overline{u} = K e^{\int_0^t r(\tau) d\tau} \left(\frac{K}{\|u_0\|_\infty} - 1 + e^{\int_0^t r(\tau) d\tau} \right)^{-1},
$$

is the solution of

$$
\begin{cases}
\frac{d\overline{u}}{dt} = r(t)\overline{u}(1 - \frac{\overline{u}}{K}), \ t > 0, \\
\overline{u}(0) = \|u_0\|_\infty.
\end{cases}
\tag{7.2.6}
$$

It follows from Theorem 7.2.1 that $x = h(t)$ is monotonically increasing. We then have that $\lim_{t \to +\infty} h(t) := h_\infty \in (0, +\infty]$.

Definition 7.2.4 The information is vanishing if

$$
h_\infty < \infty, \quad \lim_{t \to +\infty} \|u(t, .)\|_{C([0, h(t)]} = 0,
$$

while the information is spreading if

$$
h_\infty = \infty, \quad \lim_{t \to +\infty} u(t, x) = K,
$$

uniformly for x in any bounded set of $[0, \infty)$.

A direct application of Lemma 7.2.2 leads to the following lemma and corollary.

Lemma 7.2.5 Suppose that (A) hold.

(i) If $h_\infty < \infty$, then $h_\infty \leq \frac{\pi}{2}\sqrt{\frac{d}{r_\infty}}$.

(ii) If $h_\infty < \frac{\pi}{2}\sqrt{\frac{d}{r_\infty}}$, then $\lim_{t \to +\infty} \|u(t, .)\|_{C([0, h(t)])} = 0$.

Corollary 7.2.6 If $h_0 \geq \frac{\pi}{2}\sqrt{\frac{d}{r_\infty}}$, then $h_\infty = \infty$.

Theorem 7.2.7 If $h_\infty < \infty$, then $\lim_{t \to +\infty} \|u(t, x)\|_{C([0, h(t)])} = 0$.

Proof: Suppose $\limsup_{t\to+\infty}\|u(t,.)\|_{C([0,h(t)])}=\varepsilon>0$ by contradiction. We first show that for any $0<\alpha<1$

$$\|u\|_{C^{(1+\alpha)/2,1+\alpha}([0,\infty)\times[0,h(t)])}+\|h\|_{C^{1+\alpha/2}([0,\infty))}\leq C, \qquad (7.2.7)$$

where C depends on h_0, α, $\|u_0\|_{C^2([0,h_0])}$ and h_∞.

In fact, we consider a transformation $y=\frac{h_0 x}{h(t)}$, which straightens the free boundary $x=h(t)$ to the line $y=h_0$.

Let $u(t,x)=v(t,y)$, and direct calculations show that

$$u_t=v_t+v_y\frac{\partial y}{\partial t}=v_t-\frac{h'(t)}{h(t)}yv_y, \qquad u_{xx}=\frac{h_0^2}{h^2(t)}v_{yy}.$$

Hence the free boundary problem (7.2.1) becomes

$$\begin{cases} v_t-\frac{h'(t)y}{h(t)}v_y-d\frac{h_0^2}{h^2(t)}v_{yy}=r(t)v(1-\frac{v}{K}), & t>0,\ 0<y<h_0, \\ v_y(t,0)=v(t,h_0)=0, & t>0, \\ v(0,y)=u_0(y), & 0\leq y\leq h_0. \end{cases} \qquad (7.2.8)$$

It follows from Theorem 7.2.1 that

$$\|r(t)v(1-\frac{v}{K})\|_{L^\infty}\leq C_1, \quad \|\frac{h'(t)y}{h(t)}\|_{L^\infty}\leq C_2, \quad \|\frac{h_0^2}{h^2(t)}\|_{L^\infty}\leq C_3,$$

where C_1, C_2, C_3 are constants.

Using the L^p estimates for parabolic equation and Sobolev imbedding theorem gives that

$$\|v\|_{C^{(1+\alpha)/2,1+\alpha}([0,\infty)\times[0,h_0])}\leq C_4,$$

where C_4 is a constant depending on α, h_0, C_1, C_2, C_3, and $\|u_0\|_{C^2([0,h_0])}$. We immediately obtain (7.2.7).

It follows from the assumption $\limsup_{t\to+\infty}\|u(t,\cdot)\|_{C([0,h(t)])}=\varepsilon>0$ that there exists a sequence (t_k,x_k) in $(0,+\infty)\times[0,h(t))$, such that $u(t_k,x_k)\geq\varepsilon/2$ for $k\in N$, and $t_k\to+\infty$ as $k\to+\infty$. Since x_k is bounded, there exists a subsequence $\{x_{k_n}\}$ such that $x_{k_n}\to x_0\in[0,h_\infty)$ as $n\to+\infty$.

Define $u_n(t,x)=u(t_{k_n}+t,x)$ for $t\in(-t_{k_n},\infty),x\in[0,h(t_{k_n}+t))$. It follows from the parabolic regularity that $\{u_n\}$ has a subsequence $\{u_{n_i}\}$ satisfying $u_{n_i}\to\widetilde{u}$ as $i\to+\infty$.

Moreover $\widetilde{u}$ satisfies the following problem:

$$\widetilde{u}_t-d\widetilde{u}_{xx}=r_\infty\widetilde{u}(1-\frac{\widetilde{u}}{K}), \quad t\in R,\ 0<x<h_\infty.$$

Since that $\widetilde{u}(0,x_0)\geq\varepsilon/2$, we then have $\widetilde{u}>0$ in $(-\infty,\infty)\times[0,h_\infty)$. Note that $r_\infty(1-\frac{\widetilde{u}}{K})$ is bounded by $N:=\|r_\infty(1-\frac{\widetilde{u}}{K})\|_{L^\infty}$. Using the Hopf lemma to the equation $\tilde{u}_t - d\tilde{u}_{xx} \geq -N\tilde{u}$ at the point $(0, h_\infty)$ yields that $\tilde{u}_x(0, h_\infty) \leq -\sigma_0$, where σ_0 is a positive constant.

On the other hand, since $h_\infty < \infty$ and $h'(t) > 0$, we have $h'(t) \to 0$ as $t \to +\infty$. Therefore,

$$u_x(t_{k_n}, h(t_{k_n})) = -\frac{1}{\mu}h'(t_{k_n}) \to 0, \quad n \to \infty.$$

But the fact $\|u\|_{C^{(1+\alpha)/2, 1+\alpha}([0,\infty) \times [0, h(t)])} \leq C$ implies

$$u_x(t_{k_n}, h(t_{k_n})) = (u_n)_x(0, h(t_{k_n})) \to \tilde{u}_x(0, h_\infty) \leq -\sigma_0, \quad n \to \infty,$$

which leads to a contradiction. □

Lemma 7.2.8 Suppose that (A) hold. If $h_\infty = \infty$, then $\lim_{t \to +\infty} u(t, x) = K$ uniformly for x in any bounded set of $[0, \infty)$.

Proof: It follows from Corollary 7.2.3 that $u(t, x) \leq \bar{u}(t)$ for $t > 0$, $0 \leq x \leq h(t)$, where

$$\bar{u} = Ke^{\int_0^{h_0} r(\tau)d\tau}\left(\frac{K}{\|u_0\|_\infty} - 1 + e^{\int_0^{h_0} r(\tau)d\tau}\right)^{-1}$$

is the solution of the problem (7.2.6).

Clearly we obtain $\lim_{t \to +\infty} \bar{u}(t) = K$, and then $\limsup_{t \to +\infty} u(t, x) \leq K$ uniformly for $x \in [0, \infty)$. Since $h_\infty = \infty$, there exists $t_l > 0$ such that $l := h(t_l) > \frac{\pi}{2}\sqrt{\frac{d}{r_\infty}}$. Let $\underline{u}_l$ be the solution of the following problem:

$$\begin{cases} (\underline{u}_l)_t - d(\underline{u}_l)_{xx} = r_\infty \underline{u}_l(1 - \frac{\underline{u}_l}{K}), & t > t_l, \ 0 < x < l, \\ (\underline{u}_l)_x(t, 0) = 0, \underline{u}_l(t, l) = 0, & t \geq t_l, \\ \underline{u}_l(t_l, x) = u(t_l, x), & 0 \leq x \leq l. \end{cases} \quad (7.2.9)$$

Using Lemma 7.2.2 gives that $\underline{u}_l(t, x) \leq u(t, x)$ for $t \geq t_l$ and $0 \leq x \leq l$. Since $r_\infty > (\frac{\pi}{2l})^2 d$, it follows from a well-known result that $\underline{u}_l(t, l) \to \underline{u}_l^*(x)$ as $t \to +\infty$ uniformly in the compact subset of $[0, l)$, where $\underline{u}_l^*$ is the solution of the following problem:

$$\begin{cases} -d(\underline{u}_l^*)_{xx} = r_\infty \underline{u}_l^*(1 - \frac{\underline{u}_l^*}{K}), & -l < x < l, \\ \underline{u}_l^*(-l) = \underline{u}_l^*(l) = 0. \end{cases} \quad (7.2.10)$$

Applying Lemma 2.2 of [28], we know $\underline{u}_l^* \to K$ as $l \to +\infty$ uniformly in any compact subset of $[0, \infty)$. So $\liminf_{t \to +\infty} u(t, x) \geq K$ and then $\lim_{t \to +\infty} u(t, x) = K$ uniformly in any compact subset of $[0, \infty)$. □

In [27], a threshold value of μ was constructed for the species spreading or vanishing. But here we consider the effect of the initial users' size and let $u_0(x) = \lambda\varphi(x)$ with $\varphi \in \Sigma(h_0)$. It follows from Corollary 7.2.6 that if $h_0 \geq \frac{\pi}{2}\sqrt{\frac{d}{r_\infty}}$, then $h_\infty = \infty$ for any $\lambda > 0$, the other cases will be given in the following lemma.

The following lemma can be proved by use of the comparison results. Further details can be found in [67].

Lemma 7.2.9 Suppose that (A) hold.

(i) Assume that $h_0 < \frac{\pi}{2}\sqrt{\frac{d}{r(0)}}$. If λ is sufficiently small, then $h_\infty < \infty$ and $\lim_{t\to+\infty} \|u(t,x)\|_{C([0,h(t)])} = 0$.

(ii) Assume that $h_0 < \frac{\pi}{2}\sqrt{\frac{d}{r_\infty}}$. If λ is big enough, then $h_\infty = \infty$ and $\lim_{t\to+\infty} u(t,x) = K$ uniformly in any compact subset of $[0,\infty)$.

We now are ready to prove a result on threshold of λ.

Theorem 7.2.10 (Threshold Result) Suppose that (A) hold. Assuming that (u,h) is the solution of (7.2.1) with the initial value $u_0(x) = \lambda\varphi(x)$ for some $\lambda > 0$. Then there exists a $\lambda^* = \lambda^*(h_0,\varphi) \in [0,\infty)$ such that vanishing happens when $0 < \lambda \le \lambda^*$, and spreading happens when $\lambda > \lambda^*$.

Proof: It follows from Corollary 7.2.6 that spreading always happens if $h_0 \ge \frac{\pi}{2}\sqrt{\frac{d}{r_\infty}}$. Hence in this case we have $\lambda^*(h_0,\varphi) = 0$ for any φ.

For the remaining case $h_0 < \frac{\pi}{2}\sqrt{\frac{d}{r_\infty}}$, define

$$\lambda^* = \sup\{\lambda : h_\infty(\lambda\varphi) < +\infty\} \in [0,+\infty].$$

In the view of Lemma 4.1(ii), the spreading must happen when λ is large enough. Thus we get $\lambda^* \in [0,+\infty)$. It follows Lemma 7.2.5 that $h_\infty \le \frac{\pi}{2}\sqrt{\frac{d}{r_\infty}}$ when $0 < \lambda < \lambda^*$ and $h_\infty = \infty$ when $\lambda > \lambda^*$.

It remains to show that $h_\infty \le \frac{\pi}{2}\sqrt{\frac{d}{r_\infty}}$ if $\lambda = \lambda^*$. Next we use the contradiction argument to demonstrate the conclusion. Suppose that $h_\infty = \infty$ when $\lambda = \lambda^*$. Hence we can find $T > 0$ such that $h(T) > \frac{\pi}{2}\sqrt{\frac{d}{r_\infty}} + 1$. Since the solution of (7.2.1) continuously depends on the initial value. We can find a sufficiently small $\varepsilon > 0$ such that the solution with $u_0 = (\lambda^* - \varepsilon)\varphi$, denoted by $(u_\varepsilon, h_\varepsilon)$, satisfying $h_\varepsilon(T) > \frac{\pi}{2}\sqrt{\frac{d}{r_\infty}}$. This implies that the information spreading happens to the case $\lambda = \lambda^* - \varepsilon$, which contradicts to the definition of λ^*.   □

We now discuss the asymptotic spreading speed. We will show that when the information spreading happens, the free boundary moves at a constant speed for large time, that is, $\lim_{t\to+\infty} \frac{h(t)}{t} = k_0$, where k_0 is determined by the corresponding elliptic problem of (7.2.1) and satisfies $k_0 = \mu V'_{k_0}(0)$ with $V_k(x)$ satisfying

$$\begin{cases} -dV'' + kV' = r_\infty V(1 - \frac{V}{K}), & x > 0, \\ V(0) = 0. \end{cases} \tag{7.2.11}$$

The next result shows that k_0 is well-defined, and the proof is similar to the Proposition 4.1 in [27].

Proposition 7.2.11 For any $0 < k < 2\sqrt{r_\infty d}$, problem (7.2.11) admits a unique positive solution $V = V_k$. In addition, $V_k' > 0$ for $x \geq 0$, $V_{k_1}'(0) > V_{k_2}'(0)$, $V_{k_1}(x) > V_{k_2}(x)$, for $x > 0$ and $k_1 < k_2$. And for each μ there exists a unique $k_0 = k_0(\mu) \in (0, 2\sqrt{r_\infty d})$ such that $\mu V_{k_0}'(0) = k_0$.

Lemma 7.2.12 Suppose that (A) hold. If $h_\infty = +\infty$, then $\liminf_{t \to +\infty} \frac{h(t)}{t} \geq k_0$.

Proof: First we consider

$$\begin{cases} -d\omega'' + k_0\omega' = r_\infty\omega(1 - \frac{\omega}{K}), & 0 < x < l, \\ \omega(0) = \omega(l) = 0. \end{cases} \tag{7.2.12}$$

Then we define

$$\omega_0(x) = \begin{cases} \omega_{l_0}(x), & 0 \leq x \leq a_0, \\ \omega_{l_0}(a_0), & x > a_0, \end{cases}$$

where $a_0 \in (0, l_0)$, such that $\omega_{l_0}(a_0) = \max_{[0, l_0]} \omega_{l_0}$. We can verify that ω_0 satisfies

$$\begin{cases} -d\omega_0'' + k_0\omega_0' \leq r_\infty\omega_0(1 - \frac{\omega_0}{K}), & 0 \leq x < +\infty, \\ \omega_0(0) = 0, \ \omega_0(x) < K, & x \geq 0. \end{cases} \tag{7.2.13}$$

According to Lemma 7.2.8, we know that for any $\varepsilon > 0$ there exists $T := T_{\varepsilon, a_0}$ such that

$$h(T) > a_0, \ u(T, x) \geq K\sqrt{1 - \varepsilon} \quad \forall x \in [0, a_0].$$

We define

$$w = \sqrt{1 - \varepsilon}\,\omega_0(\eta(t) - x), \ t > 0, \ 0 \leq x \leq \eta(t), \ \eta(t) = (1 - \varepsilon)k_0 t + a_0, \ t > 0.$$

Direct computation gives

$$w_t - dw_{xx} \leq \sqrt{1 - \varepsilon}(k_0\omega_0' - d\omega_0'')$$
$$\leq r_\infty\sqrt{1 - \varepsilon}\,\omega_0(1 - \frac{\omega_0}{K})$$
$$\leq r_\infty w(1 - \frac{w}{K})$$

for $t > 0$ and $0 < x < \eta(t)$. Following the proof of Theorem 4.2 in [27], we get

$$\eta'(t) = (1 - \varepsilon)k_0 < \sqrt{1 - \varepsilon}\mu\omega_{l_0}'(0) = \sqrt{1 - \varepsilon}\mu\omega_0'(0) = -\mu w_x(t, \eta(t)),$$

$$w(0, x) = \sqrt{1 - \varepsilon}\,\omega_0(a_0 - x) \leq \sqrt{1 - \varepsilon}K \leq u(T, x) \quad \forall x \in [0, a_0],$$

$$w_x(t, 0) = -\sqrt{1 - \varepsilon}\,\omega_0'(\eta(t)) = 0.$$

Hence, it follows from Lemma 7.2.2 that

$$h(t+T) \geq \eta(t) \quad \text{for } t > 0.$$

Then

$$\liminf_{t\to+\infty} \frac{h(t)}{t} \geq \lim_{t\to+\infty} \frac{\eta(t-T)}{t} = (1-\varepsilon)k_0.$$

As ε is an arbitrarily positive number, we have $\liminf_{t\to+\infty} \frac{h(t)}{t} \geq k_0$. $\square$

Lemma 7.2.13 Suppose that (A) hold. If $h_\infty = +\infty$, then $\limsup_{t\to+\infty} \frac{h(t)}{t} \leq k_0$.

Proof: For any sufficiently small ϵ_0, since $\lim_{t\to+\infty} r(t) = r_\infty$, there exists $T > 0$ such that

$$r(t) < r_\infty + \epsilon_0, \quad t \geq T.$$

From the proof of Lemma 7.2.8, we know $\limsup_{t\to+\infty} u(t,x) \leq K$, uniformly for $0 \leq x < \infty$. Hence, there exists $T_0 > T$ such that

$$u(t,x) \leq K(1-\epsilon_0)^{-1}, \quad t \geq T_0, x \geq 0.$$

Following we apply the similar methods with Theorem 4.2 in [27]. Then we construct a suitable upper solution of problem (7.2.1). We first consider the following problem with the solution ν_{ϵ_0}:

$$\begin{cases} -d\nu'' + k(\epsilon_0)\nu' = (r_\infty + \epsilon_0)\nu(1 + \delta(\epsilon_0) - \frac{\nu}{K}), & x > 0, \\ \nu(0) = 0, \end{cases} \quad (7.2.14)$$

where $k(\epsilon_0) = \mu\nu'_{\epsilon_0}(0)$. If $\epsilon_0 \to 0$, then $\delta(\epsilon_0) \to 0$. With Proposition 7.2.11, $\nu'_{\epsilon_0}(x) > 0$ for $x \geq 0$, and $\nu_{\epsilon_0} \to K(1+\delta(\epsilon_0))$, as $x \to +\infty$. Hence there exists X which is a positive large number, such that

$$\nu_{\epsilon_0}(x) > K(1+\delta(\epsilon_0))(1-\epsilon_0) \text{ for } x \geq X.$$

We define

$$\overline{h}(t) = (1-\epsilon_0)^{-2}k_0 t + X + h(T_0), \quad t \geq T_0,$$

$$\overline{u}(t,x) = (1-\epsilon_0)^{-2}\nu_{\epsilon_0}(\overline{h}(t) - x), \quad t \geq T_0, 0 \leq x \leq \overline{h}(t).$$

Clearly, we obtain

$$\overline{u}(t,\overline{h}(t)) = 0, \quad \overline{u}_x(t,0) \leq 0, \quad t \geq T_0,$$

$$\overline{h}'(t) = -\mu\overline{u}_x(t,\overline{h}(t)), \quad t \geq T_0.$$

$$\overline{h}(T_0) \geq h(T_0), \quad \overline{u}(T_0,x) > 0, h(T_0) < x < \overline{h}(T_0).$$

And for $0 < x < h(T_0)$,

$$\overline{u}(T_0,x) = (1-\epsilon_0)^{-2}\nu_{\epsilon_0}(\overline{h}(T_0) - x) \geq (1-\epsilon_0)^{-1}K(1+\delta(\epsilon_0)) \geq u(T_0,x).$$

Using Lemma 7.2.2, we get

$$u(t,x) \leq \overline{u}(t,x), h(t) \leq \overline{h}(t), \quad t \geq T_0, 0 < x < \overline{h}(t).$$

Then

$$\limsup_{t \to +\infty} \frac{h(t)}{t} \leq \lim_{t \to +\infty} \frac{\overline{h}(t)}{t} = k(\epsilon_0)(1 - \epsilon_0)^{-2}.$$

It is well known that ν_{ϵ_0} continuously depends on ϵ_0, that is, as $\epsilon_0 \to 0$, $\nu_{\epsilon_0} \to V_{k_0}$, and $k(\epsilon_0) \to k_0$. Let $\varepsilon \to 0$, we easily obtain

$$\limsup_{t \to +\infty} \frac{h(t)}{t} \leq k_0.$$

$\square$

Combining Lemmas 7.2.12 and 7.2.13 gives the following main result.

Theorem 7.2.14 Suppose that (A) hold. If $h_\infty = +\infty$, then $\lim_{t \to +\infty} \frac{h(t)}{t} = k_0$.

Next we show how the asymptotic spreading speed k_0 changes as the parameters in (7.2.1) vary, see Proposition 4.3 in [27].

Theorem 7.2.15 Let k_0 be the asymptotic spreading speed determined by the Proposition 7.2.11. Then we have

$$\lim_{\frac{\mu K}{d} \to +\infty} \frac{k_0}{\sqrt{r_\infty d}} = 2, \quad \lim_{\frac{\mu K}{d} \to 0} \frac{k_0}{\sqrt{r_\infty d}} \frac{d}{\mu K} = \frac{1}{\sqrt{3}}.$$

7.2.4 Discussion

In this section we show that the free boundary $x = h(t)$ is increasing. Furthermore, the information traveling either lasts forever or suspends in finite time. In addition, the influence of the initial condition of news on its spread over online social networks is discussed. Let $u_0 = \lambda\varphi$ for some φ belongs to $\Sigma(h_0)$, it was shown that if λ is sufficiently small, the information vanishing must occur. Then it was shown that there exists a threshold λ^* which is dependent on $\varphi \in \Sigma(h_0)$ such that when $\lambda > \lambda^*$, the information with the initial data $u_0 = \lambda\varphi$ travels in the whole distance. Otherwise, the information vanishing happens.

Finally, if information spreading happens, the expanding news front $x = h(t)$ moves at a constant speed k_0 for long time. We also show that the following relation holds:

$$\lim_{\frac{\mu K}{d} \to \infty} \frac{k_0}{\sqrt{r_\infty d}} = 2. \tag{7.2.15}$$

Equation (7.2.15) indicates that the asymptotic traveling speed k_0 is close to $2\sqrt{r_\infty d}$, which is also called the minimum speed of (7.2.1) for the Fisher's equation as we shall discuss in depth in Sect. 7.5. The asymptotic traveling speeds of news fronts from free boundary problems and the minimum speeds from traveling wave solutions in Sect. 7.5 can provide a theoretical guide for how to maximize or control information propagation in online social networks. Several free boundary problems related to (5.7.1) remain to be mathematically studied. For example, information diffusion with multiple channels can give rise to a partial differential equation (6.6.1) defined in more complex domains.

7.3 Stability and Bifurcation

7.3.1 A More General Boundary Condition

In the previous chapters we assume that no information is exchanging at the ends of the interval $[l, L]$ and that the Neumann boundary condition $(I_x(l,t) = I_x(L,t) = 0)$ is plausible. The Neumann boundary condition is reasonable at the left end because no news is traveling in the left part.

Since the shortest friendship hops are used as a distance metric, the Neumann boundary condition at the right end is motivated by the theory of six degrees of separation in many social networks that most of nodes are six or fewer steps away from any other nodes. In many cases there may be substantial information exchanging at the right end, even though the groups at the right end have many fewer online users. As a result, the Robin boundary condition is more plausible to describe the flow of information at the right end.

In this section, we study stability and bifurcation of the following problem with indefinite weight and Robin boundary conditions arising from online social networks:

$$\begin{cases} I_t - (a(x)I_x)_x = \lambda r(t)I\left(h(x) - \frac{I}{K}\right), & t > 1,\ l < x < L, \\ I(1, x) = \phi(x), & l \leq x \leq L, \\ I_x(t, l) = 0,\ I_x(t, L) + \beta I(t, L) = 0, & t > 1, \end{cases} \qquad (7.3.1)$$

where $a(x) \in C^{1+\alpha}([l, L]), 0 < \alpha \leq 1$, is a positive function, $\phi(x) \in C^{2+\alpha}([l, L])$ is nonnegative and not identical to zero, and $r(t)$ is a positive Hölder-continuous function satisfying $\lim_{t\to+\infty} r(t) = r_\infty > 0$, h is a Hölder-continuous function and positive on a subinterval of $[l, L]$. $\phi \leq Kh_\infty$, where $h_\infty = \max_{x \in [l,L]} |h(x)|$. The parameter λ can be interpreted as a scale or factor of $r(t)$. The Robin boundary condition at $x = L$ reflects the fact that there is an exchange of information at the boundary. For $\beta > 0$, it indicates the flux $-I_x(t, L)$ is positive and therefore information flows to the right.

Note that problem (7.3.1) is nonautonomous with respect to time t. We refer to [65, 66, 77, 102, 119] and their references for further studies. In addition, the fact that h is an indefinite weight also introduces some essential challenges. We will be interested in investigating the bifurcation phenomena of (7.3.1) and find the bifurcation point of either information spreading or vanishing in online social networks.

7.3.2 Upper and Lower Solutions

First we consider the following eigenvalue problem:

$$\begin{cases} -(a(x)u')' = \lambda h(x)u, \ l < x < L, \\ u'(l) = 0, \ u'(L) + \beta u(L) = 0, \end{cases} \tag{7.3.2}$$

where λ is a parameter. λ_* is called an eigenvalue of (7.3.2) if there exists a nontrivial solution of (7.3.2) with $\lambda = \lambda_*$.

Lemma 7.3.1 Problem (7.3.2) admits a positive principal eigenvalue λ_1^+, and it is a simple eigenvalue. Moreover, any eigenfunction associated to λ_1^+ is positive on $[l, L]$.

This lemma can be proved by contradiction. We omit the details here.

Remark 7.3.2 Note that the complete spectral picture has been obtained in [48]. However, Lemma 7.3.1 is better than the corresponding one in [48] because any eigenfunction associated to λ_1^+ is positive on $[l, L]$. Moreover, by an argument similar to Lemma 7.3.1, we can show that any nonnegative solution of (7.3.4) is positive on $[l, L]$.

Let $X = \{u \in C^1[l, L] : u'(l) = 0, \ u'(L) + \beta u(L) = 0\}$ with the standard norm $\|\cdot\|$. Consider

$$\begin{cases} -(a(x)u')' = \lambda u \left(h(x) - \frac{u}{K}\right), \ l < x < L, \\ u'(l) = 0, \ u'(L) + \beta u(L) = 0. \end{cases} \tag{7.3.3}$$

Lemma 7.3.3 Problem (7.3.3) has an unbounded branch $\mathcal{C}$ in $\mathbb{R} \times X$ of positive solutions bifurcating from $(\lambda_1^+, 0)$. Moreover,

(i) $\mathcal{C}$ is bounded from above in u, bounded away from zero in λ, and extends to $+\infty$ in λ;
(ii) $\mathcal{C}$ is contained in the set of (λ, u) with $\lambda \geq \lambda_1^+$;
(iii) For each $\lambda > \lambda_1^+$ the positive solution with $(\lambda, u) \in \mathcal{C}$ is the unique positive solution and $\mathcal{C}$ is a smooth arc; i.e., $u(\lambda)$ depends differentiably on λ for $\lambda > \lambda_1^+$.

Proof Let $g(x, s) = h(x) - s/K$. Then we have $g(x, s) < 0$ for $s > h_\infty K$, $g(x, s) < g(x, 0)$ for $s > 0$ and $g_s(x, s) < 0$ for all $s \geq 0$. Applying Corollary 3.14 of [17] to (7.3.3), we can obtain (i)–(iii). $\square$

Clearly, the steady state solution equation of (7.3.1) is

$$\begin{cases} -(a(x)u')' = \lambda r_\infty u \left(h(x) - \frac{u}{K} \right), & l < x < L, \\ u'(l) = 0, \ u'(L) + \beta u(L) = 0, \end{cases} \tag{7.3.4}$$

where $u(x) = \lim_{t \to +\infty} I(t, x)$. The results of Lemma 7.3.3 imply that there exists a unique positive solution of (7.3.4) if $\lambda > \lambda_1^+/r_\infty$; and there is only a trivial solution of (7.3.4) if $\lambda \leq \lambda_1^+/r_\infty$.

Next we give the following definition of upper and lower solutions of (7.3.1).

Definition 7.3.4 A function $I \in C^{2,1}((l, L) \times (1, +\infty)) \cap C^{1,0}([l, L] \times [1, +\infty))$ is called an upper solution of (7.3.1) if it satisfies

$$\begin{cases} I_t - (a(x)I')' \geq f(t, x, I, \lambda), & l < x < L, \ t > 1 \\ I(1, x) \geq \phi(x), & l \leq x \leq L, \\ I_x(t, l) \geq 0, \ I_x(t, L) + \beta I(t, L) \geq 0, & t > 1, \end{cases} \tag{7.3.5}$$

where $f(t, x, I, \lambda) = \lambda r(t)I \left(h(x) - \frac{I}{K} \right)$. Similarly, I is called a lower solution of (7.3.1) if it satisfies all the reversed inequalities in (7.3.5).

Lemma 7.3.5 Let $I(t, x)$ be a solution of (7.3.1), $\widetilde{I}(t, x)$ and $\widehat{I}(t, x)$ are upper and lower solutions of (7.3.1), respectively, then $\widehat{I}(t, x) \leq I(t, x) \leq \widetilde{I}(t, x)$ in $[l, L] \times [1, +\infty)$.

Proof Let $w = \widetilde{I} - \widehat{I}$. Then by Definition 7.3.4 and the mean value theorem,

$$\begin{cases} w_t - (a(x)w')' \geq f_I(t, x, \widehat{\eta}(t, x), \lambda)w, & l < x < L, \ t > 1 \\ w(1, x) \geq 0, & l \leq x \leq L, \\ w_x(t, l) \geq 0, \ w_x(t, L) + \beta w(t, L) \geq 0, & t > 1, \end{cases}$$

where $\widehat{\eta}$ is an intermediate value between $\widetilde{I}$ and $\widehat{I}$. By Lemma 2.2.1 of [94], we have $\widetilde{I} \geq \widehat{I}$. Since I may be considered as a lower solution or an upper solution the relation $\widehat{I}(t, x) \leq I(t, x) \leq \widetilde{I}(t, x)$ follows immediately. $\square$

Finally, we prove two lemmas which will be used later.

Lemma 7.3.6 For any $\epsilon > 0$, let $u_\epsilon(x)$ be any positive solution of

$$\begin{cases} -(a(x)v')' = \lambda r_\infty \left(h(x) + \epsilon \right) v - \lambda \left(r_\infty - \epsilon \right) \frac{v^2}{K}, & l < x < L, \\ v'(l) = 0, \ v'(L) + \beta v(L) = 0. \end{cases} \tag{7.3.6}$$

Then $\liminf_{\epsilon \to 0+} u_\epsilon(x) = u(x)$, where $u(x)$ is the unique positive of (7.3.6) with $\epsilon = 0$.

Proof Consider

$$\begin{cases} -(a(x)v')' = \lambda r_\infty h(x)v - \lambda r_\infty \frac{v^2}{K}, \ l < x < L, \\ v'(l) = 0, \ v'(L) + \beta v(L) = 0. \end{cases} \tag{7.3.7}$$

Clearly, 0 is a lower solution of (7.3.7). It is easy to show that u_ϵ is an upper solution of (7.3.7). So we have that $0 \le u(x) \le u_\epsilon(x)$ for any $x \in [l, L]$. Let $w = \liminf_{\epsilon \to 0+} u_\epsilon(x)$. From (7.3.6), we get

$$\begin{cases} -(a(x)w')' = \lambda r_\infty h(x)w - \lambda r_\infty \frac{w^2}{K}, \ l < x < L, \\ w'(l) = 0, \ w'(L) + \beta w(L) = 0. \end{cases}$$

It follows that $w \equiv u$. $\square$

Similarly to the proof of Lemma 7.3.6, we can obtain the following result.

Lemma 7.3.7 For any $\epsilon > 0$, let $u_\epsilon(x)$ be any positive solution of

$$\begin{cases} -(a(x)v')' = \lambda r_\infty (h(x) - \epsilon) v - \lambda (r_\infty + \epsilon) \frac{v^2}{K}, \ l < x < L, \\ v'(l) = 0, \ v'(L) + \beta v(L) = 0. \end{cases} \tag{7.3.8}$$

Then $\limsup_{\epsilon \to 0+} u_\epsilon(x) = u(x)$, where $u(x)$ is the unique positive of (7.3.8) with $\epsilon = 0$.

7.3.3 Eigenvalue Problems

We shall get some asymptotically stable results about the solutions of (7.3.1). Firstly, consider the following auxiliary problem:

$$\begin{cases} I_t = (a(x)I')' + I(bh(x) - cI), & l < x < L, \ t > 1 \\ I(1, x) = \phi(x), & l \le x \le L, \\ I_x(t, l) = 0, \ I_x(t, L) + \beta I(t, L) = 0, & t > 1, \end{cases} \tag{7.3.9}$$

where b, c are positive constants, $\phi(x) \le bh_\infty/c$ for any $x \in [l, L]$. Consider the following eigenvalue problem:

$$\begin{cases} -(a(x)u')' = \lambda u + bh(x)u, \ l < x < L, \\ u'(l) = 0, \ u'(L) + \beta u(L) = 0. \end{cases} \tag{7.3.10}$$

Let $\Lambda(bh)$ be the first eigenvalue of (7.3.10). It is well known that $\Lambda(bh)$ is a simple, principal eigenvalue. We write φ_1 for the positive eigenfunction associated to $\Lambda(bh)$, normalized such that $\|\varphi_1\|_\infty = 1$. Moreover, we also have the following lemma.

Lemma 7.3.8 (7.3.10) has a positive solution if and only if $\lambda = \Lambda(bh)$.

Proof Let v be any eigenfunction associated to an eigenvalue $\mu > \Lambda(bh)$. We need to show that v changes sign. Assume by contradiction that $v \geq 0$, the case $v \leq 0$ being completely analogous. By the Gronwall inequality, we can easily show that $v > 0$ on $[l, L]$. By the following Picone's identity (see [4, 59]):

$$|\varphi_1'|^2 - \left(\frac{\varphi_1^2}{v}\right)' v' = |\varphi_1'|^2 + \frac{\varphi_1^2}{v^2}|v'|^2 - 2\frac{\varphi_1}{v}\varphi_1'v' \geq 0,$$

we can easily get the following identity:

$$a|\varphi_1'|^2 - \left(\frac{\varphi_1^2}{v}\right)' av' = a|\varphi_1'|^2 + \frac{a\varphi_1^2}{v^2}|v'|^2 - 2\frac{a\varphi_1}{v}\varphi_1'v' \geq 0.$$

By an easy calculation and using above identity, we obtain that

$$0 \leq \int_l^L (\Lambda(bh) - \mu)\,\varphi_1^2\,dx < 0,$$

which is impossible. Hence we have proved that v must change sign. □

Using Lemma 7.3.8, we can get the following result.

Lemma 7.3.9 If h_2 is constant, then

$$\Lambda(bh_1 + h_2) = \Lambda(bh_1) - h_2.$$

Proof Let φ_2 be any positive eigenfunction associated to $\Lambda(bh_1 + h_2)$. Then we have

$$-(a(x)\varphi_2')' = \Lambda(bh_1 + h_2)\,\varphi_2 + (bh_1 + h_2)\,\varphi_2 = (\Lambda(bh_1 + h_2) + h_2)\,\varphi_2 + bh_1\varphi_2.$$

Lemma 7.3.8 implies that

$$\Lambda(bh_1 + h_2) + h_2 = \Lambda(bh_1),$$

i.e., $\Lambda(bh_1 + h_2) = \Lambda(bh_1) - h_2$. □

Remark 7.3.10 If $a(x) \equiv 1$, Lemma 7.3.9 can obtain from relation (3.5) of [65]. However, the authors of [65] did not give the proof of relation (3.5). For reader's convenience, we give its proof here. Note that our proof can be easily applied to the corresponding high dimensional problem.

Lemma 7.3.11 Let $\Lambda(bh)$ be the first eigenvalue of (7.3.10). Then:

(i) There exists a unique positive solution of (7.3.9) denoted by $I^*(t, x)$. Moreover, $0 < I^*(t, x) \leq bh_\infty/c$ for any $x \in [l, L]$ and $t > 1$.

(ii) If $\Lambda(bh) > 0$, then (7.3.9) admits only one nonnegative steady state solution $u = 0$, which is globally asymptotically stable, that is $\lim_{t \to +\infty} I^*(t, x) = 0$ in $C^1[l, L]$. Moreover the convergence is exponentially fast and uniform for bounded sets of initial data ϕ.

(iii) If $\Lambda(bh) < 0$, then (7.3.9) has only one positive steady state solution $u = u^*(x)$, which is globally asymptotically stable, that is $\lim_{t \to +\infty} I^*(t, x) = u^*(x)$ in $C^1[l, L]$. Moreover the convergence is exponentially fast and uniform for bounded sets of initial data ϕ.

This lemma can be proved by combining a variational approach with the above results.

Remark 7.3.12 If $a(x) \equiv 1$, Lemma 7.3.11 is just the corollary of Theorem 3.3 of [65]. However, our proof is more simple and direct even in the case of $a \equiv 1$.

By Lemma 7.3.11, we can easily get the following results.

Proposition 7.3.13 Let λ_1^+ be the positive principal eigenvalue of (7.3.2).

(i) If $b < \lambda_1^+$, then (7.3.9) admits only one nonnegative steady state solution $u = 0$, which is globally asymptotically stable, that is $\lim_{t \to +\infty} I^*(t, x) = 0$ uniformly on $[l, L]$.
(ii) If $b > \lambda_1^+$, then (7.3.9) has only one positive steady state solution $u = u^*(x)$, which is globally asymptotically stable, that is $\lim_{t \to +\infty} I^*(t, x) = u^*(x)$ uniformly on $[l, L]$.

7.3.4 Information Spreading and Vanishing

The following two theorems are our main results.

Theorem 7.3.14 If $\lambda < \lambda_1^+/r_\infty$, then (7.3.1) no positive equilibria and all solutions decay to zero uniformly on $[l, L]$ as $t \to +\infty$.

Proof Clearly, 0 is a lower solution of (7.3.1). Now, consider the following problem:

$$\begin{cases} I_t - (a(x)I')' = \lambda r(t)I\left(h(x) - \frac{I}{K}\right), & l < x < L, \ t > 1 \\ I(1, x) = M\phi_1(x), & l \le x \le L, \\ I_x(t, l) = 0, \ I_x(t, L) + \beta I(t, L) = 0, & t > 1, \end{cases} \quad (7.3.11)$$

where $\phi_1 > 0$ is the eigenfunction corresponding to λ_1^+ and $M > 0$ is a constant. We choose M so large that $M\phi_1(x) \ge \phi(x)$. For any solution $\widetilde{I}(t, x)$ of (7.3.11), we can see that $\widetilde{I}$ is an upper solution of (7.3.1). Let I be any solution of (7.3.1). It follows from Lemma 7.3.5 that

$$0 \le I(t, x) \le \widetilde{I}(t, x) \text{ for any } x \in [l, L] \text{ and } t \ge 1.$$

Since $\lim_{t \to +\infty} r(t) = r_\infty$, for any $\epsilon > 0$, there exists a $\widetilde{T}_0 > 1$, such that $r_\infty - \epsilon \le r(t) \le r_\infty + \epsilon$ for $t \ge \widetilde{T}_0$. Since $\lim_{t \to +\infty} r(t)h(x) = r_\infty h(x)$, for above

ϵ, there exists a $\widehat{T}_0 > 1$, such that $r_\infty h(x) - \epsilon r_\infty \leq r(t)h(x) \leq r_\infty h(x) + \epsilon r_\infty$ for $t \geq \widehat{T}_0$ and $x \in [l, L]$. Let $T_0 = \max\left\{\widetilde{T}_0, \widehat{T}_0\right\}$. Then $\widetilde{I}(t, x)$ satisfies that

$$\widetilde{I}_t \leq \left(a(x)\widetilde{I}'\right)' + \lambda r_\infty \left(h(x) + \epsilon\right)\widetilde{I} - \lambda \left(r_\infty - \epsilon\right)\frac{\widetilde{I}^2}{K}, \quad l < x < L, \ t > T_0.$$

Now consider the following problem:

$$\begin{cases} I_t - (a(x)I')' = \lambda r_\infty \left(h(x) + \epsilon\right) I - \lambda \left(r_\infty - \epsilon\right)\frac{I^2}{K}, & l < x < L, \ t > T_0 \\ I(T_0, x) = \widetilde{I}(T_0, x), & l \leq x \leq L, \\ I_x(t, l) = 0, \ I_x(t, L) + \beta I(t, L) = 0, & t > T_0. \end{cases}$$
$$(7.3.12)$$

For the unique positive solution $\overline{I}_\epsilon(t, x)$ of (7.3.12), by the comparison principle again, we can see that

$$\widetilde{I}(t, x) \leq \overline{I}_\epsilon(t, x) \text{ for any } x \in [l, L] \text{ and } t \geq T_0.$$

On the other hand, consider the following eigenvalue problem:

$$\begin{cases} -(a(x)u')' - \lambda r_\infty h(x)u = \mu u + \epsilon \lambda r_\infty u, \ l < x < L, \\ u'(l) = 0, \ u'(L) + \beta u(L) = 0. \end{cases}$$

By Lemma 7.3.9, we have that $\Lambda \left(\lambda r_\infty(h + \epsilon)\right) = \Lambda \left(\lambda r_\infty\right) - \epsilon \lambda r_\infty$. The fact of $\lambda < \lambda_1^+/r_\infty$ implies $\lambda r_\infty \in (0, \lambda_1^+)$. So we have that $\Lambda \left(\lambda r_\infty\right) > 0$. Choose $\epsilon > 0$ sufficiently small such that $\Lambda \left(\lambda r_\infty(h + \epsilon)\right) > 0$. By Lemma 7.3.11, we have $\overline{I}_\epsilon(t, x) \to 0$ uniformly for $x \in [l, L]$ as $t \to +\infty$. Therefore, $I(t, x) \to 0$ uniformly for $x \in [l, L]$ as $t \to +\infty$. $\square$

Theorem 7.3.15 If $\lambda > \lambda_1^+/r_\infty$, then (7.3.1) has a unique positive steady state solution u^*, and all solutions to (7.3.1) satisfies $I(t, x) \to u^*(x)$ uniformly on $[l, L]$ as $t \to +\infty$.

Proof In view of the proof of Theorem 7.3.14, $\lambda > \lambda_1^+/r_\infty$ implies that $\Lambda \left(\lambda r_\infty\right) < 0$. It follows that $\Lambda \left(\lambda r_\infty(h + \epsilon)\right) < 0$. Lemma 7.3.11 implies that $\overline{I}_\epsilon(t, x) \to \overline{u}_\epsilon(x)$ uniformly for $x \in [l, L]$ as $t \to +\infty$, where $\overline{u}_\epsilon(x)$ is the unique positive solution of

$$\begin{cases} -(a(x)u')' = \lambda r_\infty \left(h(x) + \epsilon\right) u - \lambda \left(r_\infty - \epsilon\right)\frac{u^2}{K}, \ l < x < L, \\ u'(l) = 0, \ u'(L) + \beta u(L) = 0. \end{cases}$$

So we have

$$\limsup_{t \to +\infty} I(t, x) \leq \overline{u}_\epsilon(x) \text{ for any } x \in [l, L].$$

By Lemma 7.3.6, we get that

$$\limsup_{t \to +\infty} I(t, x) \leq u^*(x) \text{ for any } x \in [l, L].$$

Since $\lim_{t\to+\infty} r(t) = r_\infty$, for any $\epsilon > 0$, there exists a $\widetilde{T}_1 > 1$, such that $r_\infty - \epsilon \le r(t) \le r_\infty + \epsilon$ for $t \ge \widetilde{T}_1$. Since $\lim_{t\to+\infty} r(t)h(x) = r_\infty h(x)$, for above ϵ, there exists a $\widehat{T}_1 > 1$, such that $r_\infty h(x) - \epsilon d \le r(t)h(x) \le r_\infty h(x) + \epsilon d$ for $t \ge \widehat{T}_1$ and $x \in [l, L]$. Let $T_1 = \max\left\{\widetilde{T}_1, \widehat{T}_1\right\}$. Now let $\underline{I}(t, x)$ be the solution of

$$
\begin{cases}
W_t - (a(x)W')' = \lambda r(t)W\left(h(x) - \frac{W}{K}\right), & l < x < L,\ t > T_1 \\
W(T_1, x) = \delta I(T_1, x), & l \le x \le L, \\
W_x(t, l) = 0,\ W_x(t, L) + \beta W(t, L) = 0, & t > T_1,
\end{cases}
$$

where $\delta \in (0, 1]$ is a positive constant, $I(t, x)$ is a solution of (7.3.1). Then we can easily show $\underline{I}(t, x) \le I(t, x)$ for $t = T_1$. Hence $\underline{I}(t, x)$ is a lower solution of (7.3.1) in $[l, L] \times [T_1, +\infty)$.

Clearly, we have that

$$
\underline{I}_t \ge \left(a(x)\underline{I}'\right)' + \lambda\left(r_\infty h(x) - \epsilon d\right)\underline{I} - \lambda\left(r_\infty + \epsilon\right)\frac{I^2}{K}, \quad l < x < L,\ t > T_1.
$$

Now consider the problem

$$
\begin{cases}
W_t - (a(x)W')' = \lambda\left(r_\infty h(x) - \epsilon d\right)W - \lambda\left(r_\infty + \epsilon\right)\frac{W^2}{K}, & l < x < L,\ t > T_1 \\
W(T_1, x) = \delta I(T_1, x), & l \le x \le L, \\
W_x(t, l) = 0,\ W_x(t, L) + \beta W(t, L) = 0, & t > T_1.
\end{cases}
$$

$$(7.3.13)$$

For the unique positive solution $\widehat{I}_\epsilon(t, x)$ of (7.3.13), by the comparison principle, we have that

$$
\underline{I}(t, x) \ge \widehat{I}_\epsilon(t, x) \text{ for any } x \in [l, L] \text{ and } t \ge T_1.
$$

On the other hand, consider the following eigenvalue problem:

$$
\begin{cases}
-(a(x)u')' - \lambda r_\infty h(x)u = \mu u - \epsilon r_\infty u, & l < x < L, \\
u'(l) = 0,\ u'(L) + \beta u(L) = 0.
\end{cases}
$$

By Lemma 7.3.9, we get that $\Lambda\left(\lambda r_\infty(h - \epsilon)\right) = \Lambda\left(\lambda r_\infty\right) + \epsilon r_\infty$. The fact of $\lambda > \lambda_1^+/r_\infty$ implies $\lambda r_\infty > \lambda_1^+$. So we have that $\Lambda\left(\lambda r_\infty\right) < 0$. Choose $\epsilon > 0$ sufficiently small such that $\Lambda\left(\lambda r_\infty(h - \epsilon)\right) < 0$. By Lemma 7.3.11, we have $\widehat{I}_\epsilon(t, x) \to \widehat{u}_\epsilon(x)$ uniformly for $x \in [l, L]$ as $t \to +\infty$, where $\widehat{u}_\epsilon(x)$ is the unique positive solution of

$$
\begin{cases}
-(a(x)u')' = \lambda r_\infty\left(h(x) - \epsilon\right)u - \lambda\left(r_\infty + \epsilon\right)\frac{u^2}{K}, & l < x < L, \\
u'(l) = 0,\ u'(L) + \beta u(L) = 0.
\end{cases}
$$

Therefore, we have that

$$
\liminf_{t\to+\infty} I(t, x) \ge \widehat{u}_\epsilon(x) \text{ for } x \in [l, L].
$$

By Lemma 7.3.7, we get that

$$\liminf_{t \to +\infty} I(t, x) \geq u^*(x) \text{ for any } x \in [l, L].$$

Therefore, we have $I(t, x) \to u^*(x)$ uniformly on $[l, L]$ as $t \to +\infty$. □

Remark 7.3.16 The results of Theorems 7.3.14 and 7.3.15 show that if $\lambda < \lambda_1^+/r_\infty$, the information vanishes in finite time; if $\lambda > \lambda_1^+/r_\infty$, the information diffusion lasts forever.

Remark 7.3.17 While [65, 66] deal with more general equations, Theorem 7.3.14 and 7.3.15 are sharper than the corresponding results in [65, 66] as we are able to identify the cut-off point (λ_1^+/r_∞) for which the solutions of (7.3.1) go to either zero or a positive static state.

7.3.5 Discussion

Information diffusion, in particular, news diffusion over social media is often time sensitive. Reaction-diffusion equations arising from online social networks involves a decaying function $r(t)$. Most of the existing research focuses on the cases that $r(t)$ is constant [17] or periodic [44]. Some researchers also study eigenvalue, stability and persistence of nonautonomous parabolic PDEs [65, 81]. Mathematical analysis of associated eigenvalue and bifurcation problems can help identify thresholds for the change of social dynamics. There are many interesting results for eigenvalues, stability, bifurcation, and persistence of the reaction-diffusion equations for positive $h(x)$ and more challenging problems remain when $h(x)$ may take negative value [17, 76]. In the context of social media, $h(x)$ may be negative for some x in particular when negative news or spam are involved as many online users will delete them. With the availability of real data from social media, we are in a position to study the challenging problems from both theoretical and practical aspects, identify conditions for stability and persistence, and equally importantly, verify the mathematical criteria through the real datasets collected from social media.

The study of spatial heterogeneity on information diffusion in social media has significant theoretical and practical implications. For example, since $h(x)$ represents the adoption rate of information for the group users whose distance away from the origin is x, the shape of $h(x)$ may contribute to locating the most influenced users or opinion leaders in social media. Other related interesting problems include maximizing the total influenced users for certain classes of $h(x)$. The issue is of interest as it has commercial potentials and social implications. Much research on this issue has emerged in recent years in efforts to design efficient algorithms for detecting opinions from the corpus of data [38]. The PDE models provide a new framework to design detection algorithms by studying mathematical properties of $h(x)$. As a result, recent theoretical developments on nonlinear partial differential equations can facilitate the research and development of important social problems.

7.4 Hopf Bifurcation of an Epidemic-Like Rumor Model

Understanding the mechanism of rumor diffusion on online social networks is important for controlling and preventing rumor propagation. Without loss of generality, we divide the users of social networks into three groups: the susceptible users (S), those who have not heard the rumor yet; the infected users (I), those who have heard the rumor and will continue to spread rumors; the rational users (Y), those who have the ability to identify the authenticity of information and stop the rumor spreading. We consider rumor on social networks to spread not only over time but also over the network' space. It is assumed that the susceptible users as well as the rational users will be subject to logistic growth. Logistic growth is widely used to model the population dynamics where the growth rate is proportional to both the existing population and the amount of available resources. In this section we follow [149] to use partial functional differential equations to study rumor propagation on social networks.

$$\begin{cases} \frac{\partial S}{\partial t} = d_0 \frac{\partial^2 S}{\partial x^2} + S(a - bS(t - \tau)) - \beta_1 SI - \gamma_1 S, \\ \frac{\partial I}{\partial t} = d_0 \frac{\partial^2 I}{\partial x^2} + \beta_1 SI - \beta_2 IY - \gamma_2 I, \\ \frac{\partial Y}{\partial t} = d_0 \frac{\partial^2 Y}{\partial x^2} + Y(d - eY) + \beta_2 IY - \gamma_3 Y, \end{cases} \tag{7.4.1}$$

for $t > 0$, $x \in \Omega = [0, L]$ with homogeneous Neumann boundary condition

$$\frac{\partial S}{\partial n}(t, x) = \frac{\partial I}{\partial n}(t, x) = \frac{\partial Y}{\partial n}(t, x) = 0, \;\; t \ge 0, x \in \partial\Omega, \tag{7.4.2}$$

and initial conditions

$$\begin{cases} S(t, x) = S_0(t, x) \ge 0, \; (t, x) \in [-\tau, 0] \times \bar{\Omega}, \\ I(0, x) = I_0(0, x) \ge 0, \; x \in \bar{\Omega}, \\ Y(0, x) = Y_0(0, x) \ge 0, \; x \in \bar{\Omega}, \end{cases} \tag{7.4.3}$$

where $S(t, x)$, $I(t, x)$, and $Y(t, x)$ represent the density of the susceptible users, the infected users and the rational users with a distance of x at time t, respectively. $S_0 \in C = C([-\tau, 0], X)$ (X is defined by $X = \{u \in W^{2,2}(0, L) : u_x(0) = u_x(L) = 0\}$) is the initial density of the susceptible users. I_0 and Y_0 are the initial densities of the infected users and the rational users. The coefficients a, b, d, e, d_0, β_1, β_2, γ_1, γ_2, γ_3, L are positive constants. a is the intrinsic rate of the susceptible users growth, d is the intrinsic rate of the rational users growth, $\frac{a}{b}$ is the most carrying capacity of the susceptible users,

$\frac{d}{e}$ is the most carrying capacity of the rational users, β_1 is the spreading rate, β_2 is the recovery rate, L is the number of communities on a social network, γ_1, γ_2, and γ_3 denote the leaving rate of the three kinds of users, respectively. $\tau \geq 0$ is a time delay. $d_0 \frac{\partial^2}{\partial x^2}$ is a diffusion term, being used to describe the impact of spatio-temporal rumor propagation on the three kinds of users with a distance of x at time t. The boundary condition in (7.4.2) implies that there are no rumors across the boundary of Ω.

7.4.1 Mathematical Analysis and Simulation

In this section, we analyze the corresponding characteristic equations and further discuss the local stability and Hopf bifurcation of system (7.4.1) with the time delay τ as the bifurcation parameter. It is easy to verify that

1. for any feasible parameters, system (7.4.1) has a trivial equilibrium point $E^0 = (0,0,0)^T$;
2. if the parameters satisfy the following condition (H_1) $a - \gamma_1 > 0$, $d - \gamma_3 > 0$, $\beta_1(a-\gamma_1) - b\gamma_2 > 0$, then system (7.4.1) has four boundary equilibrium points

$$E^1 = (\frac{a-\gamma_1}{b}, 0, 0)^T, E^2 = (0, 0, \frac{d-\gamma_3}{e})^T,$$

$$E^3 = (\frac{a-\gamma_1}{b}, 0, \frac{d-\gamma_3}{e})^T, E^4 = (\frac{\gamma_2}{\beta_1}, \frac{\beta_1(a-\gamma_1) - b\gamma_2}{\beta_1^2}, 0)^T;$$

3. if the parameters satisfy the following condition (H_2) $\beta_1[e\gamma_2 + \beta_2(d-\gamma_3)] + (a-\gamma_1)\beta_2^2 > 0$, $e\beta_1(a-\gamma_1) - b[e\gamma_2 + \beta_2(d-\gamma_3)] > 0$, $\beta_1[\beta_1(d-\gamma_3) + \beta_2(a-\gamma_1)] - b\gamma_2\beta_2 > 0$, then system (7.4.1) has a unique positive equilibrium point $E^* = (S^*, I^*, Y^*)^T$, where

$$S^* = \frac{\beta_1[e\gamma_2 + \beta_2(d-\gamma_3)] + (a-\gamma_1)\beta_2^2}{e\beta_1^2 + b\beta_2^2}, \quad I^* = \frac{e\beta_1(a-\gamma_1) - b[e\gamma_2 + \beta_2(d-\gamma_3)]}{e\beta_1^2 + b\beta_2^2},$$

$$Y^* = \frac{\beta_1[\beta_1(d-\gamma_3) + \beta_2(a-\gamma_1)] - b\gamma_2\beta_2}{e\beta_1^2 + b\beta_2^2}.$$

In this section, we mainly discuss the stability and Hopf bifurcation of the positive equilibrium point. Without loss of generality, let $\tilde{S} = S - S^*$, $\tilde{I} = I - I^*$, $\tilde{Y} = Y - Y^*$, and drop bars for the simplicity of notations. Then system (7.4.1) can be transformed into the following form:

$$\begin{cases} \frac{\partial S}{\partial t} = d_0 \frac{\partial^2 S}{\partial x^2} + a_{11}S - bS^* S(t-\tau) - \beta_1 S^* I - \beta_1 SI - bSS(t-\tau), \\ \frac{\partial I}{\partial t} = d_0 \frac{\partial^2 I}{\partial x^2} + \beta_1 I^* S + a_{22}I - \beta_2 I^* Y + \beta_1 SI - \beta_2 IY, \\ \frac{\partial Y}{\partial t} = d_0 \frac{\partial^2 Y}{\partial x^2} + \beta_2 Y^* I + a_{33}Y + \beta_2 IY - eY^2, \end{cases}$$

$$(7.4.4)$$

where

$$a_{11} = a - bS^* - \beta_1 I^* - \gamma_1,$$
$$a_{22} = \beta_1 S^* - \beta_2 Y^* - \gamma_2,$$
$$a_{33} = d - 2eY^* + \beta_2 I^* - \gamma_3.$$

Thus, the positive equilibrium point $E^* = (S^*, I^*, Y^*)^T$ of system (7.4.1) is transformed into the zero equilibrium point $E_0 = (0, 0, 0)^T$ of system (7.4.4).

In the following, we will analyze the stability and Hopf bifurcation of the zero equilibrium point E_0. For simplicity, we choose $\Omega = [0, \pi]$. Let

$$U(t) = (u_1(t), u_2(t), u_3(t))^T = (S(t, \cdot), I(t, \cdot), Y(t, \cdot))^T,$$

then (7.4.4) can be rewritten as an abstract differential equation in the phase space $\mathcal{C} = C([-\tau, 0], X)$ of the form

$$\dot{U} = D\Delta U(t) + L(U_t) + f(U_t), \tag{7.4.5}$$

where
$D = diag\{d_0, d_0, d_0\},$
$\Delta = diag\{\partial^2/\partial x^2, \partial^2/\partial x^2, \partial^2/\partial x^2\},$
$U_t(\theta) = U(t + \theta), -\tau \le \theta \le 0,$
$L : \mathcal{C} \to X$
and
$f : \mathcal{C} \to X$ are given, respectively, by

$$L(\varphi) = \begin{pmatrix} a_{11}\varphi_1(0) - bS^*\varphi_1(-\tau) - \beta_1 S^*\varphi_2(0) \\ \beta_1 I^*\varphi_1(0) + a_{22}\varphi_2(0) - \beta_2 I^*\varphi_3(0) \\ \beta_2 Y^*\varphi_2(0) + a_{33}\varphi_3(0) \end{pmatrix} \tag{7.4.6}$$

and

$$f(\varphi) = \begin{pmatrix} -\beta_1\varphi_1(0)\varphi_2(0) - b\varphi_1(0)\varphi_1(-\tau) \\ \beta_1\varphi_1(0)\varphi_2(0) - \beta_2\varphi_2(0)\varphi_3(0) \\ \beta_2\varphi_2(0)\varphi_3(0) - e\varphi_3^2(0) \end{pmatrix}. \tag{7.4.7}$$

For $\varphi(\theta) = U_t(\theta)$, $\varphi = (\varphi_1, \varphi_2, \varphi_3)^T \in \mathcal{C}$, the linearized system of (7.4.5) at the zero equilibrium point E_0 is

$$\dot{U} = D\Delta U(t) + L(U_t) \tag{7.4.8}$$

and its characteristic equation is

$$\lambda y - D\Delta y - L(e^{\lambda \cdot} y) = 0, \tag{7.4.9}$$

where $y \in dom(\Delta)$, and $y \ne 0, dom(\Delta) \subset X$.

From the properties of the Laplacian operator defined on the bounded domain, the operator Δ on X has the eigenvalues $-k^2, k \in N_0 \doteq \{0, 1, 2 \cdots\}$ with the relative eigenfunctions, where

$$\beta_k^1 = \begin{pmatrix} \varsigma_k \\ 0 \\ 0 \end{pmatrix}, \beta_k^2 = \begin{pmatrix} 0 \\ \varsigma_k \\ 0 \end{pmatrix}, \beta_k^3 = \begin{pmatrix} 0 \\ 0 \\ \varsigma_k \end{pmatrix}, \varsigma_k = cos(kx). \qquad (7.4.10)$$

Clearly, $(\beta_k^1, \beta_k^2, \beta_k^3)_0^\infty$ construct a basis of the phase space X. Therefore, any element y in X can be expanded as Fourier series in the following form:

$$y = \sum_{k=0}^{\infty} Y_k^T \begin{pmatrix} \beta_k^1 \\ \beta_k^2 \\ \beta_k^3 \end{pmatrix}, Y_k = \begin{pmatrix} < y, \beta_k^1 > \\ < y, \beta_k^2 > \\ < y, \beta_k^3 > \end{pmatrix}. \qquad (7.4.11)$$

By calculation

$$L(\varphi^T(\beta_k^1, \beta_k^2, \beta_k^3)^T) = L^T(\varphi)(\beta_k^1, \beta_k^2, \beta_k^3), \quad k \in N_0. \qquad (7.4.12)$$

According to (7.4.10) and (7.4.11), (7.4.9) is equivalent to

$$\sum_{k=0}^{\infty} Y_k^T \left[\lambda I_3 + Dk^2 - \begin{pmatrix} a_{11} - bS^* e^{-\lambda\tau} & -\beta_1 S^* & 0 \\ \beta_1 I^* & a_{22} & -\beta_2 I^* \\ 0 & \beta_2 Y^* & a_{33} \end{pmatrix} \right] \times \begin{pmatrix} \beta_k^1 \\ \beta_k^2 \\ \beta_k^3 \end{pmatrix} = 0.$$
$$(7.4.13)$$

Thus the eigenvalue equation is

$$\lambda^3 + (3d_0 k^2 + eY^*)\lambda^2 + [3d_0^2 k^4 + 2eY^* d_0 k^2 + (\beta_1^2 S^* I^* + \beta_2^2 I^* Y^*)]\lambda$$
$$+[d_0^3 k^6 + eY^* d_0^2 k^4 + (\beta_1^2 S^* I^* + \beta_2^2 I^* Y^*)d_0 k^2 + e\beta_1^2 S^* I^* Y^*]$$

$$+[\lambda^2 + (2d_0 k^2 + eY^*)\lambda + (d_0^2 k^4 + eY^* d_0 k^2 + \beta_2^2 I^* Y^*)]bS^* e^{-\lambda\tau} = 0. \quad (7.4.14)$$

We now state a stability result based on the Routh–Hurwitz criteria.

Theorem 7.4.1 If (H_2) holds, then system (7.4.4) is locally asymptotically stable at the zero equilibrium point E_0 as $\tau = 0$.

Proof When $\tau = 0$, the eigenvalue equation (7.4.14) has the following form:

$$\lambda^3 + B_1\lambda^2 + B_2\lambda + B_3 = 0, \qquad (7.4.15)$$

where
$$B_1 = 3d_0 k^2 + bS^* + eY^*,$$
$$B_2 = 3d_0^2 k^4 + 2(bS^* + eY^*)d_0 k^2 + \beta_1^2 S^* I^* + \beta_2^2 I^* Y^* + beS^* Y^*,$$
$$B_3 = d_0^3 k^6 + (bS^* + eY^*)d_0^2 k^4 + (\beta_2^2 I^* Y^* + \beta_1^2 S^* I^* + beS^* Y^*)d_0 k^2$$
$$+e\beta_1^2 S^* I^* Y^* + b\beta_2^2 S^* I^* Y^*.$$

Obviously, for $\forall k \in N_0$, $\lambda = 0$ is not a root of Eq.(7.4.15). Furthermore, it is easy to show that

$$B_1 > 0, \quad B_2 > 0, \quad B_3 > 0,$$

and

$$B_1 B_2 - B_3 = 8d_0^3 k^6 + 8(bS^* + eY^*)d_0^2 k^4 + [2(\beta_1^2 S^* I^* + \beta_2^2 I^* Y^*) + 2(bS^* + eY^*)^2$$

$$+ 2beS^* Y^*]d_0 k^2 + bS^* S^* (\beta_1^2 I^* + beY^*) + eY^* Y^* (\beta_2^2 I^* + beS^*) > 0.$$

According to the Routh–Hurwitz criteria, all the roots of (7.4.15) have negative real parts. Therefore, for $\tau = 0$, the zero equilibrium point E_0 is locally asymptotically stable.

$\square$

Now we discuss the effect of the delay τ on the stability of the zero equilibrium point of system (7.4.4). Assume that $i\omega$ $(\omega > 0)$ is a root of (7.4.14). Then ω should satisfy the following equation for $k \in N_0$:

$$-i\omega^3 - (3d_0 k^2 + eY^*)\omega^2 + i\omega[3d_0^2 k^4 + 2eY^* d_0 k^2 + (\beta_1^2 S^* I^* + \beta_2^2 I^* Y^*)]$$

$$+ [d_0^3 k^6 + eY^* d_0^2 k^4 + (\beta_1^2 S^* I^* + \beta_2^2 I^* Y^*)d_0 k^2 + e\beta_1^2 S^* I^* Y^*] + [-\omega^2 + i\omega(2d_0 k^2$$

$$+ eY^*) + (d_0^2 k^4 + eY^* d_0 k^2 + \beta_2^2 I^* Y^*)]bS^*(\cos\omega\tau - i\sin\omega\tau) = 0, \quad (7.4.16)$$

which implies that

$$\begin{cases} (-\omega^2 + d_0^2 k^4 + eY^* d_0 k^2 + \beta_2^2 I^* Y^*)bS^* \cos\omega\tau + (2d_0 k^2 + eY^*)bS^* \omega \sin\omega\tau \\ = (3d_0 k^2 + eY^*)\omega^2 - [d_0^3 k^6 + eY^* d_0^2 k^4 + (\beta_1^2 S^* I^* + \beta_2^2 I^* Y^*)d_0 k^2 + e\beta_1^2 S^* I^* Y^*], \\ (2d_0 k^2 + eY^*)bS^* \omega \cos\omega\tau - (-\omega^2 + d_0^2 k^4 + eY^* d_0 k^2 + \beta_2^2 I^* Y^*)bS^* \sin\omega\tau \\ = \omega^3 - \omega(3d_0^2 k^4 + 2eY^* d_0 k^2 + \beta_1^2 S^* I^* + \beta_2^2 I^* Y^*). \end{cases}$$

$$(7.4.17)$$

Squaring and adding the two equations of (7.4.17), we derive that

$$\omega^6 + P_{1k}\omega^4 + P_{2k}\omega^2 + P_{3k} = 0, \quad (7.4.18)$$

where

$$P_{1k} = 3d_0^2 k^4 + 2eY^* d_0 k^2 + [(eY^* + bS^*)(eY^* - bS^*) - 2I^*(\beta_1^2 S^* + \beta_2^2 Y^*)],$$

$$P_{2k} = 3d_0^4 k^8 + 4eY^* d_0^3 k^6 + 2(eY^* + bS^*)(eY^* - bS^*)d_0^2 k^4$$

$$+ [2b^2 \beta_2^2 S^{*2} I^* Y^* - b^2 e^2 S^{*2} Y^{*2} + I^{*2}(\beta_1^2 S^* + \beta_2^2 Y^*)^2 - 2e^2 \beta_1^2 S^* I^* Y^{*2}]$$

$$+ 2eY^*(-b^2 S^{*2} + \beta_2^2 I^* Y^* - 2\beta_1^2 S^* I^*)d_0 k^2,$$

$$P_{3k} = [3d_0^3 k^6 + (bS^* + eY^*)d_0^2 k^4 + (\beta_2^2 I^* Y^* + \beta_1^2 S^* I^* + beS^* Y^*)d_0 k^2$$

$$+ S^* I^* Y^* (e\beta_1^2 + b\beta_2^2)] \cdot [3d_0^3 k^6$$

$$+ (eY^* - bS^*)d_0^2 k^4 + (\beta_2^2 I^* Y^* + \beta_1^2 S^* I^* - beS^* Y^*)d_0 k^2 + S^* I^* Y^* (e\beta_1^2 - b\beta_2^2)].$$

Let $z = \omega^2$. Then Eq. (7.4.18) becomes

$$z^3 + P_{1k}z^2 + P_{2k}z + P_{3k} = 0. \tag{7.4.19}$$

For simplicity, denote

$$h(z) = z^3 + P_{10}z^2 + P_{20}z + P_{30}. \tag{7.4.20}$$

Now we have the following lemma.

Lemma 7.4.2 If (H_3) $e\beta_1^2 - b\beta_2^2 < 0$ holds, then Eq. (7.4.19) has at least one positive root for $k = 0$.

Under the condition of Lemma 7.4.2, without loss of generality, we assume that
(7.4.19) has three positive roots, defined by z_1, z_2, z_3, respectively. Then we have

$$\omega_1 = \sqrt{z_1}, \ \omega_2 = \sqrt{z_2}, \ \omega_3 = \sqrt{z_3}.$$

According to Eq. (7.4.17), we have

$$\cos \omega_l \tau_l = -\frac{e\beta_1^2 \beta_2^2 I^{*2} Y^{*2}}{b[(\omega_l^2 - \beta_2^2 I^* Y^*)^2 + e^2 Y^{*2}\omega_l^2]},$$

thus, if we denote

$$\tau_l^i = \frac{1}{\omega_l}\left[\arccos\left(-\frac{e\beta_1^2 \beta_2^2 I^{*2} Y^{*2}}{b\left[(\omega_l^2 - \beta_2^2 I^* Y^*)^2 + e^2 Y^{*2}\omega_l^2\right]}\right) + 2j\pi\right], \tag{7.4.21}$$

where $l = 1, 2, 3; j = 0, 1, 2, \ldots$, then $\pm i\omega_l$ is a pair of purely imaginary roots of (7.4.18) with τ_l^j. Furthermore, we can easily prove that $\pm i\omega_l$ is also a pair of simple purely imaginary roots.

Thus, we can define

$$\tau_0 = \tau_{l_0}^0 = \min_{1 \le l \le 3}\{\tau_l^0\}, \quad \omega_0 = \omega_{l_0}. \tag{7.4.22}$$

For further discussion, we make the following assumptions:

(H_4) $d_0^2 \ge max\{e^2 Y^{*2}, \ b^2 S^{*2} + 2\beta_1^2 S^* I^*, \ b^2 S^{*2} + 2I^*(\beta_1^2 S^* + \beta_2^2 Y^*) - e^2 Y^{*2}\}$

(H_5) $d_0^2(d_0 + eY^* - bS^*) - bS^*Y^*(ed_0 + \beta_2^2 I^*) > 0.$

Lemma 7.4.3 If the conditions $(H_4) \sim (H_5)$ satisfy. Then, for $\forall k \ge 1$, Eq.(7.4.18) has no positive real roots.

Proof Clearly, we can calculate that $P_{1k} > 0, P_{2k} > 0, P_{3k} > 0$, Since $P_{1k} > 0$, $P_{2k} > 0$, $P_{3k} > 0$ for $\forall k \ge 1$, we can easily obtain that Eq.(7.4.19) has no positive roots when $k \ge 1$. Thus, Eq.(7.4.18) also has no any positive roots for $\forall k \ge 1$.

Lemma 7.4.4 Let $\lambda(\tau) = \gamma(\tau) \pm i\omega(\tau)$ be the root of (7.4.14) near $\tau = \tau_0$ satisfying $\gamma(\tau_0) = 0$, $\omega(\tau_0) = \omega_0$. Assume further that $h'(\omega_0^2) > 0$, where $h(\omega_0^2)$ is defined by (7.4.20). Then, the following transversality condition holds:

$$\frac{d(\operatorname{Re}\lambda(\tau))}{d\tau}\bigg|_{\tau=\tau_0} > 0.$$

Proof When $k = 0$, differentiating the two sides of (7.4.14) with respect to τ yields

$$\left(\frac{d\lambda}{d\tau}\right)^{-1} = \frac{[3\lambda^2 + 2eY^*\lambda + I^*(\beta_1^2 S^* + \beta_2^2 Y^*)]e^{\lambda\tau} + bS^*(2\lambda + eY^*)}{\lambda bS^*(\lambda^2 + eY^*\lambda + \beta_2^2 I^* Y^*)} - \frac{\tau}{\lambda}.$$

Therefore, by $h'(\omega_0^2) > 0$, we can easily obtain

$$\left[\frac{d(\operatorname{Re}\lambda(\tau))}{d\tau}\right]^{-1}_{\tau=\tau_0, \lambda=i\omega_0} = \operatorname{Re}\left[\frac{[3\lambda^2 + 2eY^*\lambda + I^*(\beta_1^2 S^* + \beta_2^2 Y^*)]e^{\lambda\tau}}{bS^*\lambda(\lambda^2 + eY^*\lambda + \beta_2^2 I^* Y^*)}\right]_{\tau=\tau_0, \lambda=i\omega_0}$$

$$= \frac{h'(\omega_0^2)}{b^2 S^{*2}[(\omega_0^2 - \beta_2^2 I^* Y^*)^2 + e^2 Y^{*2}\omega_0^2]} > 0.$$

$\square$

Applying Theorem 7.4.1 and Lemmas 7.4.2–7.4.4, it is easy to obtain the following conclusion.

Theorem 7.4.5 Let τ_0 be defined by (7.4.22). If the parameters satisfy the conditions $(H_2) \sim (H_5)$. Suppose furthermore that the condition $h'(\omega_0^2) > 0$ holds. Then the following statements are true.

(i) When $\tau \in [0, \tau_0)$, the zero equilibrium point E_0 of system (7.4.4) is locally asymptotically stable;

(ii) The Hopf bifurcation occurs at $\tau = \tau_0$. That is, system (7.4.4) has a branch of periodic solutions bifurcating from the zero equilibrium point E_0 near $\tau = \tau_0$.

As an example, we examine the impact of delay τ on the density of the users on social networks. Let $a = 0.3$, $b = 0.5$, $d = 0.2$, $e = 0.3$, $\beta_1 = 0.45$, $\beta_2 = 0.35$, $\gamma_1 = 0.1$, $\gamma_2 = 0.1$, $\gamma_3 = 0.15$, $d_0 = 2$. By calculating, the positive equilibrium point $E^* = (0.3760, 0.0266, 0.1977)^T$. It is easy to show that $\tau_0 = 7.9189$. The positive equilibrium point E^* is locally asymptotically stable when $\tau = 3 < \tau_0$, and unstable when $\tau = 10 > \tau_0$. Furthermore, according to Theorem 7.4.5, delay can change stability. Let the parameters be the same as the above. Figure 7.5 gives the maximum and minimum densities of the infected users for different delays, where the red line represents the maximum density and the blue line represents the minimum density. From it we can find that when $\tau < 8$ the maximum density equals the minimum density, which implies E^* is locally asymptotically stable. That is close to the theoretical value $\tau = 7.9189$.

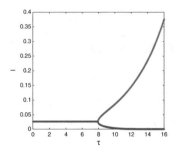

Fig. 7.5 Maximum and minimum densities of the infected users for different time delays

7.4.2 Discussion

In this section, we discuss theoretical analysis and numerical simulations of how time delay affects rumor propagation at the mean-field level. Because of the rapid development of mobile social networks, the traditional rumor spreading models, which only consider the information spreads only over time, cannot accurately describe the spatial dynamic characteristics of rumor spreading. Based on partial functional differential equations, the spatio-temporal rumor spreading model of social networks perfectly accounts for the deficiencies of previous literature. To describe the impact of time delay, the local stability of the positive equilibria point are investigated by taking the delay τ as the bifurcation parameter. The theoretical analysis reveals that the delay τ can effectively change the stability of the system. Finally, numerical simulations are conducted to study the properties of the rumor spreading.

7.5 Traveling Wave Solutions and Spreading Speeds

7.5.1 Long-Term Propagation of Information on Social Networks

In general, news diffusion in social media is time-sensitive and the influence of news decays drastically as time elapses. However, some information can take a longer period to spread in social media. Cha et al. [19] examined the aggregate growth patterns of two sets of 5346 and 897 photos in the Flickr social network that are older than 1 year and 2 years, respectively. It is found in [19] that the long-term trend in the number of cumulative fans exhibits a pattern of steady linear growth. Many photos show quick rise in popularity during the first few days after being loaded. However, most of the pictures exhibit a period of steady linear growth after the first few (10–20) days. More importantly the linear growth is sustained over an extended period. The growth rate continues to increase even after 1 or 2 years. Thus, for long-term propagation of information, we may choose distance metrics in a way that online users can be embedded in the whole x-axis and the source of information can be viewed from either from $-\infty$ or ∞. Furthermore, the parameters may be chosen to be independent of time t. As such, it is meaningful to discuss long-time behavior and traveling solutions of the reaction-diffusion systems for information diffusion in online social networks. A traveling wave solution often represents a transition process connecting two steady states of interactive populations. Traveling wave fronts of partial differential equations are solutions of the form $u(x + ct)$ that have a fixed shape and translate at a constant speed c as time evolves. The wave speed c can be interpreted as the rate of spread of the introduced population in biology and information in online social networks. The theoretical results on traveling wave solution of reaction-diffusion equations have successfully predicted spread rates of some introduced species.

7.5.2 Mathematical Formulation and Assumptions

For the long-time behavior and spatial spread of an advantageous gene in a population, Fisher [31] and Kolmogorov et al. [58] studied the nonlinear parabolic equation

$$u_t = du_{xx} + f(u) \qquad (7.5.1)$$

here, $u(x, t)$ represents the population density at location x and time t and $f(0) = f(1) = 0$ and $f(u) > 0$ with no Allen effort. Traveling wave fronts of (7.5.1) are of interest since they enable us to better understand how a population propagates. It was shown that (7.5.1) has a traveling wave solution of the form $u(x + ct)$ if and only if $|c| \geq c^*$ and the minimum speed of propagation for (7.5.1) is c^* where

$$c^* = 2\sqrt{f'(0)d}$$

This basic formula $c^* = 2\sqrt{f'(0)d}$ establishes the speeding spreads for nonlinear parabolic equations and indicates that the rate of spread is a linear function of time and that it can be predicted quantitatively as a function of measurable life history parameters.

   In spatial biology and epidemiology, it is of great interest to estimate how fast a species or infectious disease spreads within a population. Building on the mathematical foundation for the theory of spreading speeds for cooperative systems by Weinberger et al. [135], the authors [125] discussed spreading speeds for a large class of systems of reaction-diffusion equations which are not necessarily cooperative through analysis of traveling waves via the convergence of initial data to wave solutions. In particular, [125] provides a practical approach to calculate the propagation speed based on the eigenvalues of the parameterized Jacobian matrix of its linearized system at the initial state.

In this section, we extend our work on traveling waves of reaction-diffusion systems in [125] to long-term propagation of information on social networks where underlying partial differential equations may be non-cooperative. We consider a system of reaction-diffusion equations with zero and another positive equilibria

$$u_t = Du_{xx} + f(u) \text{ for } x \in \mathbb{R}, \ t \geq 0. \tag{7.5.2}$$

with

$$u(x,0) = u_0(x) \text{ for } x \in \mathbb{R}, \tag{7.5.3}$$

where $u = (u_i)$, $D = \text{diag}(d_1, d_2, \ldots, d_N)$, $d_i > 0$ for $i = 1, \ldots, N$

$$f(u) = (f_1(u), f_2(u), \ldots, f_N(u)),$$

$u_0(x)$ is a bounded uniformly continuous function on $\mathbb{R}$. By a solution we mean a continuous function u, which is twice continuously differentiable with respect to x and once continuously differentiable with respect to t, satisfies an appropriate system of equations.

To deal with a non-cooperative system, we shall assume that there are two additional cooperative systems

$$u_t = Du_{xx} + f^+(u) \text{ for } x \in \mathbb{R}, \ t \geq 0 \tag{7.5.4}$$

$$u_t = Du_{xx} + f^-(u) \text{ for } x \in \mathbb{R}, \ t \geq 0 \tag{7.5.5}$$

where f^+ lies above and f^- below of f.

The following assumption will enable us to make use of the corresponding results for cooperative systems in [135] to establish spreading speeds for (7.5.2).

(H1) (i) Assume that $D = \text{diag}(d_1, d_2, \ldots, d_N)$, $d_i > 0$ for $i = 1, \ldots, N$. Let $k^+ = (k_i^+) >> 0$ and $f : [0, k^+] \to \mathbb{R}^N$ to be a continuous and twice piecewise continuously differentiable function. Assume that $\mathcal{C}_{k^+}$ is an invariant set of (7.5.2) in the sense that for any given $u_0 \in \mathcal{C}_{k^+}$, the solution of (7.5.2) with the initial condition u_0 exists and remains in $\mathcal{C}_{k^+}$ for $t \in [0, \infty)$.

(ii) Let $0 << k^- = (k_i^-) \leq k = (k_i) \leq k^+$. Assume there exist continuous and twice piecewise continuously differentiable function $f^\pm = (f_i^\pm) : [0, k^+] \to \mathbb{R}^N$ such that for $u \in [0, k^+]$

$$f^-(u) \leq f(u) \leq f^+(u).$$

(iii) $f(0) = f(k) = 0$ and there is no other positive equilibrium of f between 0 and k. $f^\pm(0) = f^\pm(k^\pm) = 0$. There is no other positive equilibrium of $f^\pm$ between 0 and $k^\pm$. f has finite number of equilibria.

(iv) Equations (7.5.4) and (7.5.5) are cooperative (i.e. $\partial_i f_j^{\pm}(u) \geq 0$ for $u \in [0, k^{\pm}], i \neq j$).

(v) $f^{\pm}(u), f(u)$ have the same Jacobian matrix $f'(0)$ at $u = 0$.

A traveling wave solution u of (7.5.2) is a solution of the form $u = u(x + ct), u \in C(\mathbb{R}, \mathbb{R}^N)$. Substituting $u(x,t) = u(x + ct)$ into (7.5.2) and letting $\xi = x + ct$, we obtain the wave equation

$$Du''(\xi) - cu'(\xi) + f(u(\xi)) = 0 \text{ for } \xi \in \mathbb{R}. \tag{7.5.6}$$

Now if we look for a solution of the form $(u_i) = (e^{\lambda\xi}\eta_\lambda^i), \lambda > 0, \eta_\lambda = (\eta_\lambda^i) >> 0$ for the linearization of (7.5.6) at the origin, we arrive at the following system:

$$\operatorname{diag}(d_i\lambda^2 - c\lambda)\eta_\lambda + f'(0)\eta_\lambda = 0$$

which can be rewritten as the following eigenvalue problem:

$$\frac{1}{\lambda}A_\lambda\eta_\lambda = c\eta_\lambda, \tag{7.5.7}$$

where

$$A_\lambda = (a_\lambda^{i,j}) = \operatorname{diag}(d_i\lambda^2) + f'(0)$$

The matrix $f'(0)$ has nonnegative off-diagonal elements. In fact, there is a constant α such that $f'(0) + \alpha I$ has nonnegative entries, where I is the identity matrix. By reordering the coordinates, we can assume that $f'(0)$ is in block lower triangular form, in which all the diagonal blocks are irreducible or 1 by 1 zero matrix. A matrix is irreducible if it is not similar to a lower triangular block matrix with two blocks via a permutation. From the Perron–Frobenius theorem any irreducible matrix A with nonnegative entries has a unique principal positive eigenvalue (which is the spectral radius of A, $\rho(A)$) with a corresponding principal eigenvector with strictly positive coordinates. For an irreducible matrix A with nonnegative off-diagonal elements, we shall call the eigenvalue $\rho(A+\alpha I)-\alpha$ of A, which has the same positive eigenvector, the principal eigenvalue of A (see e.g. [135]). Let

$$\Psi(A) = \rho(A + \alpha I) - \alpha.$$

Here $A + \alpha I$ is irreducible and nonnegative, and $\rho(A + \alpha I)$ is the spectral radius of $A + \alpha I$.

(H2)   Assume that A_λ with irreducible blocks is in block lower triangular form. Further assume that its first diagonal block has the positive principal eigenvalue $\Psi(A_\lambda)$, and $\Psi(A_\lambda)$ is strictly larger than the principal eigenvalues of all other irreducible diagonal blocks for $\lambda > 0$. In addition, assume that there is a positive eigenvector $\nu_\lambda = (\nu_\lambda^i) >> 0$ of A_λ corresponding to $\Psi(A_\lambda)$, and that ν_λ is continuous with respect to λ for $\lambda > 0$.

Let

$$\Phi(\lambda) = \frac{1}{\lambda}\Psi(A_\lambda) > 0.$$

Now we state Lemma 7.5.1, which shall enable us to calculate the minimum speed and give accurate asymptotic estimates of traveling solutions. Its proof can be found in [125]. When Φ is a strictly convex function of λ, it clearly satisfies Lemma 7.5.1.

Lemma 7.5.1 Assume that $(H1)$–$(H2)$ hold. Then

(1) $\Phi(\lambda) \to \infty$ as $\lambda \to 0$;
(2) $\Phi(\lambda) \to \infty$ as $\lambda \to \infty$;
(3) $\Phi(\lambda)$ is decreasing near $\lambda = 0$ and $\lambda > 0$;
(4) $\Psi(A_\lambda)$ is a convex function of λ for $\lambda > 0$;
(5) $(\lambda^2\Phi'(\lambda))' = \lambda\frac{d^2\Psi(A_\lambda)}{d\lambda^2} \geq 0$;
(6) $\Phi'(\lambda)$ changes sign at most once on $(0,\infty)$
(7) $\Phi(\lambda)$ assumes its minimum

$$c^* = \inf_{\lambda>0} \Phi(\lambda) > 0 \qquad\qquad (7.5.8)$$

at a finite λ.
(8) For each $c > c^*$, there exist $\Lambda_c > 0$ and $\gamma \in (1,2)$ such that

$$\Phi(\Lambda_c) = c, \quad \Phi(\gamma\Lambda_c) < c.$$

That is

$$\frac{1}{\Lambda_c}A_{\Lambda_c}\nu_{\Lambda_c} = \Phi(\Lambda_c)\nu_{\Lambda_c} = c\nu_{\Lambda_c}$$

and

$$\frac{1}{\gamma\Lambda_c}A_{\gamma\Lambda_c}\nu_{\gamma\Lambda_c} = \Phi(\gamma\Lambda_c)\nu_{\gamma\Lambda_c} < c\nu_{\gamma\Lambda_c}$$

where $\nu_{\Lambda_c} \gg 0, \nu_{\gamma\Lambda_c} \gg 0$ are positive eigenvectors of $\frac{1}{\Lambda_c}A_{\Lambda_c}, \frac{1}{\gamma\Lambda_c}A_{\gamma\Lambda_c}$ corresponding to eigenvalues $\Phi(\Lambda_c)$ and $\Phi(\gamma\Lambda_c)$, respectively.

In addition to (H1–H2), we also need assumption (H3) which only requires the nonlinearity is less than its linearization along the particular function $\nu_\lambda e^{-\lambda x}$, which means the nonlinearity does not display an Allee effect for the particular function.

(H3)   Assume that for any $\alpha > 0, \lambda > 0$

$$f^\pm(\alpha\nu_\lambda) \leq \alpha f'(0)\nu_\lambda, \quad \text{where } \nu_\lambda = (\nu_\lambda^i).$$

7.5.3 Spreading Speed and Traveling Wave Solutions

The following theorem is the main results of Sect. 7.5, which is also valid when (7.5.2) is cooperative as we can choose $f^{\pm} = f$. We will establish the asymptotic spreading speed (Theorem 7.5.2 (i–ii)) for general non-cooperative reaction-diffusion systems and further characterize the spreading speed as the speed of the slowest nonconstant traveling wave solutions (Theorem 7.5.2 (iii–v)).

Theorem 7.5.2 Assume $(H1)$–$(H3)$ hold. Then the following statements are valid:

(i) For any $u_0 \in \mathcal{C}_k$ with compact support, the solution $u(x,t)$ of (7.5.2) with (7.5.3) satisfies

$$\lim_{t\to\infty} \sup_{|x|\geq tc} u(x,t) = 0, \text{ for } c > c^*$$

(ii) For any vector $w \in \mathbb{R}^N, w \gg 0$, there is a positive R_w with the property that if $u_0 \in \mathcal{C}_k$ and $u_0 \geq w$ on an interval of length $2R_w$, then the solution $u(x,t)$ of (7.5.2) with (7.5.3) satisfies

$$k^- \leq \liminf_{t\to\infty} \inf_{|x|\leq tc} u(x,t) \leq k^+, \text{ for } 0 < c < c^*$$

(iii) For each $c > c^*$ (7.5.2) admits a traveling wave solution $u = u(x + ct)$ such that $0 \ll u(\xi) \leq k^+, \xi \in \mathbb{R}$,

$$k^- \leq \liminf_{\xi\to\infty} u(\xi) \leq \limsup_{\xi\to\infty} u(\xi) \leq k^+$$

and

$$\lim_{\xi\to-\infty} u(\xi)e^{-\Lambda_c\xi} = \nu_{\Lambda_c}. \tag{7.5.9}$$

If, in addition, (7.5.2) is cooperative in $\mathcal{C}_k$, then u is nondecreasing on $\mathbb{R}$.

(iv) For $c = c^*$ (7.5.2) admits a nonconstant traveling wave solution $u = u(x + ct)$ such that $0 \leq u(\xi) \leq k^+, \xi \in \mathbb{R}$.

(v) For $0 < c < c^*$ (7.5.2) does not admit a traveling wave solution $u = u(x + ct)$ with $\liminf_{\xi\to\infty} u(\xi) \gg 0$ and $u(-\infty) = 0$.

In order to prove the main result, we state the following comparison theorem for cooperative systems of reaction-diffusion equations which is a consequence of the maximum principle (see, e.g., Protter and Weinberger [98]).

Theorem 7.5.3 Let D be a positive definite diagonal matrix. Assume that $F = (F_j)$ is a vector-valued function in $\mathbb{R}^N$ is continuous and piecewise continuously differentiable in $\mathbb{R}$ and the underlying system is cooperative in the sense that for each j, F_j is nondecreasing in all but the jth component. Suppose that $u(x, t), v(x, t)$ satisfy

$$u_t - Du_{xx} - F(u) \leq v_t - Dv_{xx} - F(v) \tag{7.5.10}$$

If $u(x, t_0) \leq v(x, t_0)$ for $x \in \mathbb{R}$, then

$$u(x, t) \leq v(x, t) \text{ for } x \in \mathbb{R}, t \geq t_0.$$

Proof of Theorem 7.5.2 (i) and (ii).

Part (i).   For a given $u_0 \in \mathcal{C}_k$ with compact support, let $u^+(x, t)$ be the solutions of (7.5.4) with the same initial condition u_0 as the solution u of (7.5.2). Then the Comparison Principle 7.5.3 implies that $u^+(x, t) \in \mathcal{C}_{k+}$ and

$$u_t^+ - Du_{xx}^+ - f^+(u^+) = 0 = u_t - Du_{xx} - f(u) \geq u_t - Du_{xx} - f^+(u).$$

and

$$0 - D0 - f^-(0) = 0 = u_t - Du_{xx} - f(u) \leq u_t - Du_{xx} - f^-(u).$$

Therefore the Comparison Principle 7.5.3 implies that

$$0 \leq u(x, t) \leq u^+(x, t) \text{ for } x \in \mathbb{R}, t > 0.$$

Thus for any $c > c^*$, it follows from [135] that

$$\lim_{t \to \infty} \sup_{|x| \geq tc} u^+(x, t) = 0,$$

and hence

$$\lim_{t \to \infty} \sup_{|x| \geq tc} u(x, t) = 0,$$

Part (ii).   According to [135], for any strictly positive constant ω, there is a positive R_ω (choose the larger one between the R_ω for (7.5.4) and the R_ω for (7.5.5)) with the property that if $u_0 \geq \omega$ on an interval of length $2R_\omega$, then the solutions $u^\pm(x, t)$ of (7.5.4) and (7.5.5) with the same initial value u_0 are in $\mathcal{C}_{k+}$ and satisfy

$$\liminf_{t \to \infty} \inf_{|x| \leq tc} u^\pm(x, t) = k^\pm, \text{ for } 0 < c < c^*.$$

As before we have

$$u_t^+ - Du_{xx}^+ - f^+(u^+) = 0 = u_t - Du_{xx} - f(u) \geq u_t - Du_{xx} - f^+(u)$$

and

$$u^- - Du^-_{xx} - f^-(u^-) = 0 = u_t - Du_{xx} - f(u) \le u_t - Du_{xx} - f^-(u).$$

Thus, the Comparison Principle 7.5.3 implies that

$$u^-(x,t) \le u(x,t) \le u^+(x,t) \text{ for } x \in \mathbb{R}, t > 0.$$

Thus for any $c < c^*$, it follow from [135] that

$$\liminf_{t \to \infty} \inf_{|x| \le ct} u^{\pm}(x,t) = k^{\pm},$$

and hence

$$k^- \le \liminf_{t \to \infty} \inf_{|x| \le ct} u(x,t) \le k^+.$$

Proof of Theorem 7.5.2 (iii). In order to find traveling waves for (7.5.2), we will apply the Schauder's fixed point theorem.

Let $u = (u_i) \in \mathcal{A}$ and define two integral operators

$$\mathcal{T}^{\pm}[u] = (\mathcal{T}^{\pm}_i[u])$$

for f^- and f^+

$$\mathcal{T}^{\pm}_i[u](\xi)$$
$$= \frac{1}{d_i(\lambda_{1i} + \lambda_{2i})} \left[\int_{-\infty}^{\xi} e^{-\lambda_{1i}(\xi-s)} H^{\pm}_i(u(s)) ds + \int_{\xi}^{\infty} e^{\lambda_{2i}(\xi-s)} H^{\pm}_i(u(s)) ds \right]$$
$$(7.5.11)$$

and

$$H^{\pm}_i(u(s)) = \beta u_i(s) + f^{\pm}_i(u(s)).$$

It is easy to see that both $\mathcal{T}^+$ and $\mathcal{T}^-$ are monotone. In view of the fact that f^- is nondecreasing, there exists a nondecreasing fixed point $u^- = (u^-_i)$ of $\mathcal{T}^-$ such that $\mathcal{T}^-[u^-] = u^-$, $\lim_{\xi \to \infty} u^-_i(\xi) = k^-_i, i = 1, \ldots, N$, and $\lim_{\xi \to -\infty} u^-_i(\xi) = 0, i = 1, \ldots, N$. Furthermore, $\lim_{\xi \to -\infty} u^-_i(\xi) e^{-\Lambda_c \xi} = \nu^i_{\Lambda_c}$ for $i = 1, \ldots, N$. Let

$$\widetilde{\phi^+}(\xi) = (\widetilde{\phi^+_i}(\xi)),$$

where

$$\widetilde{\phi^+_i}(\xi) = \min\{k^+_i, \nu^i_{\Lambda_c} e^{\Lambda_c \xi}\} \text{ for } i = 1, \ldots, N, \ \xi \in \mathbb{R};$$

It follows that $u^-_i(\xi) \le \widetilde{\phi^+_i}$ for $\xi \in \mathbb{R}, i = 1, \ldots, N$. Now let

$$\mathcal{B} = \{u : u = (u_i) \in \mathcal{E}_\varrho, u^-(\xi) \le u(\xi) \le \widetilde{\phi^+}(\xi) \text{ for } \xi \in (-\infty, \infty)\}, \quad (7.5.12)$$

where $\mathcal{E}_\varrho$ is defined in [125]. It is clear that $\mathcal{B}$ is a bounded nonempty closed convex subset in $\mathcal{E}_\varrho$. Furthermore, we have, for any $u = (u_i) \in \mathcal{B}$

$$u_i^- = T_i^-[u^-] \leq T_i^-[u] \leq T_i[u] \leq T_i^+[u] \leq T_i^+[\widetilde{\phi^+}] \leq \widetilde{\phi_i^+} \text{ for } i = 1, \ldots, N.$$

Therefore, $\mathcal{T} : \mathcal{B} \to \mathcal{B}$. We can show that $\mathcal{T} : \mathcal{B} \to \mathcal{B}$ is continuous and maps bounded sets into compact sets. Therefore, the Schauder fixed point theorem shows that the operator $\mathcal{T}$ has a fixed point u in $\mathcal{B}$, which is a traveling wave solution of (7.5.2) for $c > c^*$. Since $u_i^-(\xi) \leq u_i(\xi) \leq \widetilde{\phi_i^+}(\xi), \xi \in (-\infty, \infty), i = 1, \ldots, N$, it is easy to see that for $i = 1, \ldots, N$, $\lim_{\xi \to -\infty} u_i(\xi) = 0$, $\lim_{\xi \to -\infty} u_i(\xi)e^{-\Lambda_c \xi} = \nu_{\Lambda_c}^i$,

$$k^- \leq \liminf_{\xi \to \infty} u(\xi) \leq \limsup_{\xi \to \infty} u(\xi) \leq k^+$$

and $0 < u_i^-(\xi) \leq u_i(\xi) \leq k_i^+$ for $\xi \in (-\infty, \infty)$.

Proof of Theorem 7.5.2 (iv). We adopt the limiting approach to prove Theorem 7.5.2 (iv). For each $n \in \mathbb{N}$, choose $c_n > c^*$ such that $\lim_{n \to \infty} c_n = c^*$. According to Theorem 7.5.2 (iii), for each c_n there is a traveling wave solution u_n of (7.5.2) such that

$$u_n = \mathcal{T}[u_n](\xi).$$

and

$$k^- \leq \liminf_{\xi \to \infty} u_n(\xi) \leq \limsup_{\xi \to \infty} u_n(\xi) \leq k^+.$$

It is easy to show that (u_n) is equicontinuous and uniformly bounded on $\mathbb{R}$, the Ascoli's theorem implies that there is a vector-valued continuous function u on $\mathbb{R}$ and subsequences (u_{n_m}) of (u_n) such that

$$\lim_{m \to \infty} u_{n_m}(\xi) = u(\xi)$$

uniformly in ξ on any compact interval of $\mathbb{R}$. Further in view of the dominated convergence theorem we have

$$u = \mathcal{T}[u](\xi)$$

Here the underlying $\lambda_{1i}, \lambda_{2i}$ of $\mathcal{T}$ are dependent on c and continuous functions of c. Thus u is a traveling solution of (7.5.2) for $c = c^*$. Because of the translation invariance of u_n, we always can assume that the first component of $u_n(0)$ equals to a sufficiently small positive number for all n. Note that there are only finite number of equilibria. Consequently u is a nonconstant traveling solution of (7.5.2).

Proof of Theorem 7.5.2 (v). Suppose, by contradiction, that for some $c \in (0, c^*)$, (7.5.2) has a traveling wave $u(x, t) = u(x + ct)$ with $\liminf_{\xi \to \infty} u(\xi) \gg 0$ and $u(-\infty) = 0$. Thus $u(x, t) = u(x + ct)$ can be larger than a positive vector with arbitrary length. It follows from Theorem 7.5.2 (ii)

$$\liminf_{t \to \infty} \inf_{|x| \leq ct} u(x, t) \geq k^- \gg 0, \text{ for } 0 < c < c^*$$

Let $\hat{c} \in (c, c^*)$ and $x = -\hat{c}t$. Then

$$\lim_{t\to\infty} u\big(-(\hat{c}-c)t\big) = \lim_{t\to\infty} u(-\hat{c}t, t) \geq \liminf_{t\to\infty} \inf_{|x|\leq t\hat{c}} u(x,t) >> 0.$$

However,

$$\lim_{t\to\infty} u\big(-(\hat{c}-c)t\big) = u(-\infty) = 0,$$

which is a contradiction.

$\square$

7.5.4 Propagation Speeds of Cooperative Information

As we discuss earlier, certain information such as photos in the Flickr social network can take a long period to spread and exhibit a pattern of steady linear growth. In this case, we can assume, that, in (6.3.1), $b_i = 0, i = 1, 2$, all other coefficients are positive constants and $h_i(x) = 1, i = 1, 2$. (6.3.1) is a cooperative system since its Jacobian

$$\begin{pmatrix} r_1 - 2\frac{r_1 u_1}{k_1} + \alpha_1 u_2 & a_1 u_1 \\ \alpha_2 u_2 & r_2 - 2\frac{r_2 u_2}{k_2} + \alpha_2 u_1 \end{pmatrix}$$

has nonnegative off-diagonal elements for $u_1, u_2 \geq 0$. Now we can use (7.5.8) to calculate the minimum speed of information propagation in social media for a longer period of time where r_i, α_i are all positive constant. We are interested in a transition process connecting two equilibria $(0, 0)$ and (e_1, e_2) of (6.3.1) where

$$e_1 = \frac{k_1 r_2(\alpha_1 k_2 + r_1)}{r_1 r_2 - \alpha_1 \alpha_2 k_1 k_2}, e_2 = \frac{k_2 r_1(\alpha_2 k_1 + r_2)}{r_1 r_2 - \alpha_1 \alpha_2 k_1 k_2}$$

We assume that

$$r_1 r_2 - \alpha_1 \alpha_2 k_1 k_2 > 0 \tag{7.5.13}$$

and therefore $e_1, e_1 > 0$. We can apply Theorems 7.5.2 to 6.3.1 to calculate the minimum speed of (6.3.1) for information propagation from $(0, 0)$ and (e_1, e_2). As a result, there is a traveling wave solution connecting $(0, 0)$ and (e_1, e_2) and the minimum speed of information propagation can be calculated by the formula (7.5.8). For simplicity, assume that and $d_1 \geq d_2$ and $r_1 \geq r_2$. Now it is easy to calculate that the Jacobian of (6.3.1) at $(0, 0)$ is

$$\begin{pmatrix} r_1 & 0 \\ 0 & r_2 \end{pmatrix}$$

For $\lambda \geq 0$, the largest eigenvalue $\Psi(A_\lambda)$ of the matrix

$$\begin{pmatrix} d_1\lambda^2 + r_1 & 0 \\ 0 & d_2\lambda^2 + r_2 \end{pmatrix}$$

is $d_1\lambda^2 + r_1$. Therefore

$$\Phi(\lambda) = \frac{1}{\lambda}\Psi(A_\lambda) = \inf_{\lambda > 0} \frac{d_1\lambda^2 + r_1}{\lambda}$$

In view of (7.5.8), a standard calculation shows the propagation speed for (6.3.1) is

$$c^* = 2\sqrt{d_1 r_1}$$

On the other hand, if $d_2 \geq d_1$ and $r_2 \geq r_1$, the propagation speed for (6.3.1) is

$$c^* = 2\sqrt{d_2 r_2}$$

This indicates that if (7.5.13) holds, or the effect of the interaction of the two sources are not too large, the propagation speed for multiple information sources is largely determined by the more popular source.

7.5.5 Propagation Speeds for Competing Information

If two pieces of information compete with each other to maximize their own influences on online social networks, (6.4.1) may be used to model the interaction. We assume that, in (6.4.1), $b_i = 0, i = 1, 2$, all other coefficient are positive constants and $h_i(x) = 1, i = 1, 2$ as we focus on its long-term behavior. Equation (6.4.1) is not a cooperative system. However, we are interested in a transition process connecting two equilibria $(k_1, 0)$ and $(0, k_2)$ of (6.4.1). Equation (6.4.1) can be brought into a cooperative system by the transformation $v_1 = u_1$ and $v_2 = k_2 - u_2$

$$\begin{aligned} \frac{\partial v_1}{\partial t} &= d_1 \frac{\partial^2 v_1}{\partial x^2} + r_1 v_1 (1 - \frac{v_1}{k_1}) - \alpha_1 v_1 (k_2 - v_2) \\ \frac{\partial v_2}{\partial t} &= d_2 \frac{\partial^2 v_2}{\partial x^2} - r_2(k_2 - v_2)\frac{v_2}{k_2} + \alpha_2 v_1 (k_2 - v_2) \end{aligned} \tag{7.5.14}$$

The Jacobian of (7.5.14)

$$\begin{pmatrix} r_1 - 2\frac{r_1 v_1}{k_1} - \alpha_1(k_2 - v_2) & \alpha_1 v_1 \\ \alpha_2(k_2 - u_2) & -r_2 + 2\frac{r_2 v_2}{k_2} - \alpha_2 v_1 \end{pmatrix}$$

has nonnegative off-diagonal elements for $v_1, v_2 \geq 0$ and $v_2 \leq k_2$. We assume that

$$r_1 > \alpha_1 k_2 \tag{7.5.15}$$

to ensure that the growth of u_1 remains even with the competition from u_2. We also assume that $d_1 \geq d_2$, information u_1 is not less popular than information u_2.

We are interested in a transition process connecting two equilibria $(0,0)$ and (k_1, k_2) of (7.5.14) that correspond to the two equilibria $(0, k_2)$ and $(k_1, 0)$ of (6.4.1). Again we can apply Theorem 7.5.2 to calculate the minimum speed of (7.5.14) for information propagation from $(0,0)$ and (k_1, k_2). As a result, there is a traveling wave solution of (7.5.14) connecting its two equilibria $(0,0)$ and (k_1, k_2) and the minimum speed of the information propagation can be calculated by the formula (7.5.8). Now it is easy to calculate that the Jacobian of (7.5.14) at $(0,0)$ is

$$\begin{pmatrix} r_1 - \alpha_1 k_2 & 0 \\ \alpha_2 k_2 & -r_2 \end{pmatrix}$$

For $\lambda \geq 0$, the largest eigenvalue $\Psi(A_\lambda)$ of the matrix

$$\begin{pmatrix} d_1 \lambda^2 + r_1 - \alpha_1 k_2 & 0 \\ \alpha_2 k_2 & d_2 \lambda^2 - r_2 \end{pmatrix}$$

is $d_1 \lambda^2 + r_1 - \alpha_1 k_2$. Therefore

$$\Phi(\lambda) = \frac{1}{\lambda} \Psi(A_\lambda) = \inf_{\lambda > 0} \frac{d_1 \lambda^2 + r_1 - \alpha_1 k_2}{\lambda}$$

In view of (7.5.8), a standard calculation shows the propagation speed for (7.5.14) is

$$c^* = 2\sqrt{d_1 (r_1 - \alpha_1 k_2)}$$

The conclusion indicates that information u_1 will win the competition if the growth of u_1 sustains even with the competition from u_2, and information u_1 is not less popular than information u_2. The propagation speed of the information is largely determined by the popularity and growth of the winner minus negative affect from the competition.

7.5.6 Propagation Speeds for Spatial Epidemiological Models

Spatial structures play an important role in describing the spreading of communicable diseases with epidemiological models. Here we would like to examine the traveling solutions and information adoption rate of a spatial SIR model for information diffusion in online social networks in a closed population consisting of susceptible individuals ($S(t)$), infected individuals ($I(t)$) (i.e. adopted the information) and removed individuals ($R(t)$) (i.e. refractory). The diffusive SIR model with the standard incidence takes the following form:

$$\partial_t S = d_1 \partial_{xx} S - \beta SI/(S+I)$$
$$\partial_t I = d_2 \partial_{xx} I + \beta SI/(S+I) - \gamma I \qquad (7.5.16)$$
$$\partial_t R = d_3 \partial_{xx} R + \gamma I$$

here γ is the removal rate of the infected group, β is the adoption (or influence) rate between the susceptible and infectious groups. $d_1, d_2, d_3 > 0$ represent the popularity of information with each of the groups. In general, the information is more popular for the I group than the S group. Therefore, it is assumed that $d_2 \geq d_1$. For the long-term propagation of information in online social networks, it is understood that the adoption rate β can be constant.

Equation (7.5.16) is an extension of the SI model (6.5.1) with the refractory group R. Equation (7.5.16) is a non-cooperative system. In general it still is an open question to show it is linearly determinate. Nevertheless, for (7.5.16), we can show c^* is the cut-off point of the existence of traveling wave solutions. A traveling wave solution of (7.5.16) with the form $(S(x+ct), I(x+ct,t), R(x+ct,t))$ represents the transition process of information diffusion from the initial adoption-free equilibrium $(S_{-\infty}, 0, R_{-\infty})$ to another adoption-free state $(S_\infty, 0, R_\infty)$ with S_∞ being determined by the influence rate β and the remove rate γ, as well as possibly the popularity of information. As such, it is important to determine whether traveling waves exist and what the propagation speed c is. Thus we shall look for traveling wave solutions of the form $(S(x+ct), I(x+ct), R(x+ct))$. Because R does not appear in the system of equations for the susceptible individuals S and infected individuals I, we omit the R equation and study the following system with S and I only:

$$\partial_t S = d_1 \partial_{xx} S - \beta SI/(S+I)$$
$$\partial_t I = d_2 \partial_{xx} I + \beta SI/(S+I) - \gamma I \qquad (7.5.17)$$

which satisfies the following boundary conditions at infinity:

$$S(-\infty) = S_{-\infty}, \; S(\infty) < S_{-\infty}, \; I(\pm\infty) = 0. \qquad (7.5.18)$$

In the context of infectious disease, [127] studied the traveling waves and propagation speed of (7.5.17). The result of [127] is applicable for information diffusion in online social networks. For (7.5.17), the nonlinearity f in (7.5.2) is no longer cooperative and some of the off-diagonal elements of f' may be negative. It is still an open question what additional conditions would guarantee that $\Phi(\lambda)$ maintains the convex-like property. However, for (7.5.17), $\Phi(\lambda)$ is a convex function and we note that the minimum wave speed can be obtained by its linearization at the initial state $(S_{-\infty}, 0)$. In fact, it is easy to calculate that the Jacobian of (7.5.17) at $(S_{-\infty}, 0)$ is

$$\begin{pmatrix} 0 & -\beta \\ 0 & \beta - \gamma \end{pmatrix}$$

Its largest eigenvalue is $\beta - \gamma$. For $\mu \geq 0$ and $d_2 \geq d_1$, the largest eigenvalue $\Psi(A_\lambda)$ of the matrix

$$\begin{pmatrix} d_1 \lambda^2 & -\beta \\ 0 & d_2 \lambda^2 + \beta - \gamma \end{pmatrix}$$

is $d_2 \mu^2 + \beta - \gamma$. Therefore

$$\Phi(\lambda) = \frac{1}{\lambda} \Psi(A_\lambda) = \inf_{\lambda > 0} \frac{d_2 \lambda^2 + \beta - \gamma}{\lambda}$$

In view of (7.5.8), a standard calculation shows the wave speed for (7.5.17) is

$$c^* = 2\sqrt{d_2(\beta - \gamma)}$$

In addition, [127] shows that $c^* = 2\sqrt{d_2(\beta - \gamma)}$ is the cut-off value of c for which there is a traveling wave for (7.5.17) of the form $(S(x + ct), I(x + ct))$. Specifically, it is shown in [127] that if $R_0 := \beta/\gamma > 1$ (R_0 is the basic reproduction number for the corresponding ordinary differential system) and $c > c^* := 2\sqrt{d_2(\beta - \gamma)}$, then there exists a nontrivial and nonnegative traveling wave solution (S, I) of (7.5.17) such that the boundary conditions (7.5.18) are satisfied. Furthermore, S is monotonically decreasing, $0 \leq I(x) \leq S(-\infty) - S(\infty)$ for all $x \in \mathbb{R}$, and

$$\int_{-\infty}^{\infty} \gamma I(x) dx = \int_{-\infty}^{\infty} \frac{\beta S(x) I(x)}{S(x) + I(x)} dx = c[S(-\infty) - S(\infty)]. \qquad (7.5.19)$$

On the other hand, if $R_0 = \beta/\gamma \leq 1$ or $c < c^* := 2\sqrt{d_2(\beta - \gamma)}$, then there exist no nontrivial and nonnegative traveling wave solutions (S, I) of (7.5.17) satisfying the boundary conditions (7.5.18).

$c^* = 2\sqrt{d_2(\beta - \gamma)}$ is particularly of interest as it is the cut-off point for the existence of traveling waves of (7.5.17). In other words, the cut-off speed for traveling waves of (7.5.17) is determined by its linearized systems. $c^* = 2\sqrt{d_2(\beta - \gamma)}$ can be viewed as the speed of (7.5.17) for information to spread in a social network. The result also indicates that the diffusion speed of information is proportional to the square root of the product of the popularity of the information and difference of the adoption rate and remove rate of the adopted group.

7.5.7 Discussion

In this section, we discuss mathematical properties of traveling wave solutions arising from online social networks. We use abstract mathematical theories such as fixed point theorems to show the existence of traveling wave solutions and determine wave speeds. For online social networks, the traveling wave solutions will also depend on the distance we choose. Thus we will pay particular attention to how network topology and clustering affect the spreading speed. Our approach allows us to employ techniques from not only mathematical analysis, but also from graph theory and data mining as well. Even from the mathematical point of view, there are many unsolved problems. For examples, we are also interested in competition models involving three or more competitors. In this case, showing there is a traveling wave solution is still a challenging problem.

Chapter 8
Applications

Abstract In this chapter we present two applications of partial differential equation models for information diffusion in online social networks. We present a diffusion-advection PDE model to describe a transnational diffusion process of social movement in social media during the Egyptian revolution in 2011. We develop a PDE-based influenza surveillance system by analyzing flu related Twitter data. The system aims to predict flu trends at more localized levels by leveraging the availability of geocoded Twitter data.

8.1 Introduction

The partial differential equation models provide a new efficient approach of studying social phenomena and predicting spread of news in social media. In this chapter we discuss applications of the PDE models in analysis of social movement during the Egyptian revolution in 2011 and prediction of the spread of influenza. Part of the materials in this chapter is based on the authors' two papers [63, 64] on analysis of social media during the Egyptian Revolution 2011 and [131] on modeling flu trends with real-time geo-tagged Twitter data streams.

8.2 Analysis of Twitter Information Diffusion During the Egyptian Revolution

8.2.1 The 2011 Egyptian Revolution and Social Media

The Egyptian Revolution of 2011, part of the Arab Spring, began on January 25, 2011. It involved demonstrations, marches, acts of civil disobedience,

© Springer Nature Switzerland AG 2020   113
H. Wang et al., *Modeling Information Diffusion in Online Social Networks
with Partial Differential Equations*, Surveys and Tutorials in the Applied
Mathematical Sciences 7, https://doi.org/10.1007/978-3-030-38852-2_8

strikes, occupations of plazas, riots and non-violent civil resistance. Millions of protesters from a wide range of backgrounds demanded the overthrow of Egyptian President Hosni Mubarak who had ruled Egypt since 1981. The violent clashes resulted in at least 846 deaths and more than 6000 injures. On February 11, 2011 Mubarak was forced to resign as president. On June 2, 2012 Mubarak was found guilty in the murder of protesters and sentenced to life imprisonment. He was eventually cleared of all charges on November 29, 2014. After the revolution, the Muslim Brotherhood took power in Egypt for a time. Eventually, General Abdel Fattah El-Sisi was elected president in 2014.

In response to the popularity of social media and blogs in recent years, Egyptian activism leadership turned to a decentralized body of new generational activists who took advantage of social networks to spread information and reinforce solidarity. Information diffusion in social media becomes a key factor in mobilization for success in technology-driven social movements, which have remained consistent with old-fashioned social movements. As a prerequisite for successful mobilization, effective information diffusion is essential to give salience to protest efforts. On an individual level, information spreading through online networks may easily increase "noticeability" of fellow users' participation, which would otherwise be visible only within a small local group setting. On a collective level, the increased visibility of protest efforts contributes to the accumulation of greater knowledge about the issue and the fostering of favorable attitudes communally. The aggregated attitude and knowledge then shape the climate of opinion with which individuals compare their own beliefs and decide whether to join the actions.

8.2.2 Data Collection

We first used Twitter's public streaming API to collect data with the search keyword "Egypt" between January 25th 2011 and February 20th, 2011, eight times per day, for an hour each session. Through this process, a total of 50,778 Twitter user IDs were identified. Because, as of 2011, Twitter did not allow keyword search for historical data older than 5 days, we had to access historic data by backtracking each Twitter user ID who may have tweeted around the time of the Egyptian revolution. Therefore, we took the following three steps to collect historic data: (1) identify and collect Twitter user IDs who tweeted around the time period of the Egyptian revolution, (2) backtrack all identified user IDs to retrieve all their posted content on Twitter, and (3) perform keyword searches (i.e. "Egypt") to identify relevant tweets.

The tweets are categorized into three types: ad-hoc reporting, situation verifying information, and collective action-supportive messages for exploring how different message types affect diffusion patterns. The classification was developed from collective behavior literature, which suggests that collective behavior evolves from improvised information sharing into collective identity reinforcement and strategic action mobilization. The framework was applied to social media context [93]. We identified three message types, which we summarize as follows.

- Ad-hoc reporting

  1. Definition: Provides firsthand observation without presenting any supporting materials; includes an immediate update about a situation or problem without providing additional sources.
  2. Example: "My mobile phone connection keeps getting cut every 15 s... Bad coverage or purposeful strategy?"

- Situation verifying

  1. Definition: Provides information verified or supported by a third-party, or with solid supporting evidences such a photo, video, and interviews, etc.
  2. Example: "Egypt's frustrated young wait for their lives to begin and dream of revolution http://bit.ly/iceIYZa."

- Action supportive

  1. Definition: Either of the types: (a) An emotive expression of unity with the movement (b) Sharing of action-related skills, knowledge, or tactics.
  2. Example: "My heart is with all my people in # Egypt"

The classification of message is significant for parameter selections of the PDE model. The range of the value of a parameter may be different for different type of messages. Once we perform Matlab simulations and determine the range for each type of message, we could use the PDE model with the given ranges for future prediction.

8.2.2.1 Spatial Factors: Geo-Location and International Relation

To explore diffusion patterns in a spatial dimension, we regroup tweets with their geographical information. Two types of geographic information are available in Twitter: geographic metadata and users' self-disclosed place cues. Although geographic metadata provides more accurate information, it is often restricted by Twitter's default privacy settings; users must enable the location-sharing feature on Twitter to send geo-referenced tweets. We were able to identify 82% of the identified users' geographical information.

We decided to code geographical information on a global region level instead for a few reasons. First, a fair number of users were found to be associated with multiple countries within the same regional bloc—for example, one user profile mentioned Egypt and UAE without any additional cue as to which country the user was in the moment of tweeting. Second, economic trade and migration information was only available only on a regional level. Third, some countries had a number of tweets that were too small to fit reliably with the model. While this higher-order classification might not fully capture a nation-level diffusion pattern, it increased consistency in coding and model accuracy. We categorized users' geographical information into seven global regions based on The Economist Intelligence Unit Report (2010) and the World Bank (2010): Western Europe (WE), Eastern Europe (EE), North America (NA), Latin America/the Caribbean (LA), Asia/Australasia (Asia), Middle East/North Africa (MENA), Sub-Saharan Africa (Africa). Given that some countries were ambiguously positioned between WE and EE, we decided to categorize a country as WE if the country was either a member or a potential candidate for the European Union. Since Egypt was the home country of the revolution, it was treated as a separate single entity from other MENA and was designated as the origin locational point (x_1). In total, the data contained eight spatial points.

Partial differential equation models depend on spatial arrangement of the tweets to analyze spatial patterns of information diffusion. To this end, we considered Egypt as the origin location x_1. To arrange spatial "distance" of each regional point $U(x_1, x_2, \ldots, x_8)$ from Egypt, we defined the distance in four different ways based on each international-relational factor. The first one was physical distance: a region geographically proximate to Egypt was considered to be "closer" to Egypt than those regions that were geographically far. Thus, the spatial arrangement based on geographic proximity was

$$\begin{aligned} U_{\text{proximity}}(x) &= \{Egypt, MENA, EE, WE, Africa, NA, Asia, LA\} \\ &= \{1, 2, 3, 4, 5, 6, 7, 8\} \end{aligned} \tag{8.2.1}$$

The second distance was drawn from cultural proximity, with the assumption that a larger diaspora community (i.e., from migration) would induce greater cultural proximity. The arrangement based on cultural proximity was

$$\begin{aligned} U_{\text{disapora}}(x) &= \{Egypt, MENA, WE, NA, Asia\} \\ &= \{1, 2, 3, 4\} \end{aligned} \tag{8.2.2}$$

The third distance was the volume of bilateral economic trade with Egypt that was used as criterion to define economic proximity. The arrangement based on economic proximity was

$$\begin{aligned} U_{\text{trade}}(x) &= \{Egypt, MENA, WE, NA, Asia, Africa, Asia, LA\} \\ &= \{1, 2, 3, 4, 5, 6, 7, 8\} \end{aligned} \tag{8.2.3}$$

Finally, the influence of democratic ideas was conceived as an ideological distance. The arrangement based on ideological distance was

$$U_{\text{democracy}}(x) = \{Egypt, NA, WE, LA, EE, Asia, Africa, MENA\}$$
$$= \{1, 2, 3, 4, 5, 6, 7, 8\} \tag{8.2.4}$$

meaning the more democratic regions would be more influential to (thus "closer" to) Egyptian protesters. The level of democracy for each region was based on information from the Economist Intelligence Unit Report (2010), bi-trade relations with Egypt from the Egyptian International Trade Point (2010), and emigration from Egypt from Migration Policy Centre (2009). For all criteria, the most updated data prior to the revolution were referenced. In the experiment, we will compare the ability of information spreading along the four spatial arrangements. The increment between x_i is chosen to be 1 for simplicity. Realistically, the increment between x_i can be different.

8.2.3 PDE Modeling of Global Information Diffusion

In this section, we extend the spatio-temporal modeling approach in Sect. 5.4 to incorporate an advection term to describe information diffusion. Our approach is based on deterministic dynamical equations. While conventional linear regression models and other statistical models examine correlations (or covariance) between a dependent variable and one or more independent variables, the analytic focus of diffusion modeling, including information diffusion in social media, is on a rate of change (or derivative) of one variable over the course of changes in another variable (e.g., time). This type of phenomenon can be better modeled by differential equations that equate the rate of change of one variable to that of another or their combinations. The advantage of differential equation-based modeling over linear regression models is that it allows for dynamic modeling, the main purpose of which is to understand the process of changes with respect to time. Partial differential equation models are continuous in both space and time and we assume that any change in the amount of information is restricted to one spatial dimensional line labeled as the spatial variable x. The locational (or spatial) points we constructed in this study is based on a discrete set of points (U_x) in the x-axis, which were then extended into a continuous interval.

Here we extend the diffusive logistic model by adding the "advection" term, which denotes the tendency of an object to move along with a discrete set of locations. The model is thus called a diffusive-advection logistic model, or simply a diffusion-advection model. Diffusive-advection models often describe substances of an entity (e.g., pollutants) to be carried by a bulk motion of the transport medium (e.g., fluid) [74]. For example, in modeling the spreading of infectious disease, while the local diffusion process may occur due to autonomous and random search movements of a mosquito, wind current may also result in an "advection" movement of large masses of mosquitoes and consequently cause a quick advance of infection [117].

Global information diffusion in social media may be driven by three mechanisms: the first one is diffusion, which accounts for interactions of users between different locations, the second one is advection, which may occur due to larger global geo-political forces such as democracy, migration, physical distance, and other significant forces, and the third one is local growth, by which individual local users share information relatively autonomously. Therefore, the use of diffusion-advection equations may appropriately distinguish the two mechanisms, thereby shedding light on the role of international relational forces in facilitating the spread of information during the Egyptian revolution. For formalization, suppose that a series of spatial points is arranged by a certain criterion, $U(x)$, as well as a set of time points, $T(t)$. Let $I = I(x,t)$ be the volume of information at location x and at time t. The mathematical formalization of the diffusion-advection model is:

$$\frac{\partial I}{\partial t} = \frac{\partial(de^{-bx}I_x)}{\partial x} - g(x)\frac{\partial I}{\partial x} + r(t)I(h(x) - \frac{I}{K})$$

$$I(x,1) = \phi(x), \quad l < x < L$$

$$\frac{\partial I}{\partial x}(l,t) = \frac{\partial I}{\partial x}(L,t) = 0, \quad t > 1 \tag{8.2.5}$$

$$r(t) = A + De^{-B(t-C)^2}$$

where $\frac{\partial(de^{-bx}I_x)}{\partial x}$ in Eq. (8.2.5) represents the diffusion process across $U(x)$ by unexamined forces (at random). The second part of Eq. (8.2.5), $-g(x)\frac{\partial I}{\partial x}$, is the advection term, which denotes the information shift across spatial points. The third part of (8.2.5), $r(t)I(h(x)-I/K)$, is a logistic function that captures the intrinsic growth of information over time within a particular location. This logistic function is what most ODE diffusion models are based on. Each notation is explained below:

- I: the volume of dependent variable (e.g., a volume of Twitter messages at each time and at each location)
- t: time
- x: locational (or spatial) point
- d: a parameter associated with unknown factors that may promote the spread of I
- b: a parameter associated with unknown factors that may inhibit the spread of I
- $\frac{\partial I}{\partial t}$: a rate of change in I at time t
- $\frac{\partial I}{\partial x}$: a rate of change in I at a locational point of x
- r: the intrinsic growth rate is specified as a growth function denoted by a natural exponential function (e). We assumed that the volume of I rapidly increases at the beginning and the rate of growth will decrease over time and reaches its peak at certain time
- $-g(x)$: the parameter of advection term (A coefficient that determines the rate of change in I over locational points xs. The negative sign indicates

that the shift of I is directed from one spatial point to the right-side neighbor across the x-axis)

- $h(x)$: the parameter representing the heterogeneity of intrinsic growth rate at location x. (A coefficient that determines a rate of change over time within a particular location x)
- K: the carrying capacity (the maximum possible volume of I at a given location x)
- $I(x, 1) = \phi(x)$: The initial function (the volume of tweets at time t $= 1$ to be $\phi(x)$, which specifies that the initial function has to be always ≥ 0)
- $\frac{\partial I}{\partial x}(l, t) = \frac{\partial I}{\partial x}(L, t) = 0$ for $x = l, L$, where l is the lower bound and L is the upper bound of the distance between the origin location x1 and other locations (A differential equation model requires a boundary condition. This particular formula is called the Neumann boundary condition, which assumes zero flux of I across the boundary at $x = l, L$, meaning that tweets are clustered within a single location x.)

We are in particular interested in the value of advection parameter, $g(x)$, because it represents the magnitude of information shifted across locations due to the examined force. To determine the value of $g(x)$, we used the Matlab program to best fit the empirical data to the suggested model. Under an assumption that de^{-bx} is the same or close for different types of spatial arrangement, we can compare the values of $g(x)$ and determine how global information spreads along the spatial arrangements.

To understand the effect of the advection term in diffusion-advection equations, we take a simple example and study its explicit solutions. Assuming that $d = 0, r(t) = 0$, $g(x)$ is a constant $c > 0$, then (8.2.6) becomes

$$I_t = -cI_x \tag{8.2.6}$$

in the interval $(-\infty, \infty)$. The solution of (8.2.6) is

$$I(x, t) = F(x - ct)$$

where F is a differentiable function. Such solutions of (8.2.6) are called right-traveling waves, since the graph of $F(x - ct)$ is just the graph of $F(x)$ shifted to the right ct spatial units. Figure 8.1 indicates that as time t increases, the wave of profile of $F(x)$ moves to the right, undistorted, with its shape unchanged, at speed c. For information diffusion, the advection term $g(x)$ can be interpreted as the tendency of information moving to certain direction. Assuming that the local diffusion $\frac{\partial(de^{-bx}I_x)}{\partial x}$ is the same or close for different types of spatial arrangements, we can compare the value of $g(x)$ to determine how fast information moves along a direction.

$r(t)$ in (8.2.5) reflects the increase or decay of news influence with respect to time t. In general, news in social media is time-sensitive and the influence of news stories decays drastically as time elapses. For the data on the Egyptian revolution, we examine tweets in three different periods in 2011, January 25

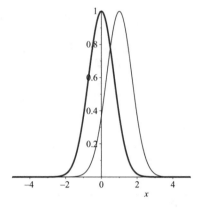

Fig. 8.1 Effect of advection term on the solution of (8.2.6)

to January 31, February 1 to February 7 and February 8 to February 13. It turns out that the daily volume of tweets first approximately increases, and then decreases. To reflect this phenomenon, we choose

$$r(t) = A + De^{-B(t-C)^2}$$

where A represents the residual rate as time increase, D, B are decay or increase rates. $r(t)$ reaches its peak at $t = C$. Figure 8.2 shows two different shapes with different values of A, B, C, D. For the simulations in the next sections, these values are obtained from a Matlab fitting program.

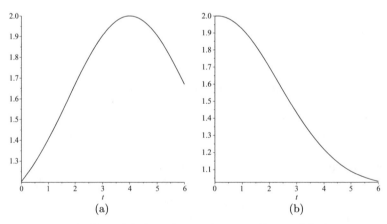

Fig. 8.2 Temporal effect. (**a**) $A = D = 1, B = 0.1, C = 4$. (**b**) $A = D = 1, B = 0.1, C = 0.1$

Model		Raw Tweet volumes			Residuals Values		
Comm Type	Distance Arranged by	1/25~1/31	2/1~ 2/7	2/8~ 2/13	1/25~1/31	2/1~ 2/7	2/8~ 2/13
Ad-hoc Reporting	Democracy	92.88%	91.72%	92.19%	92.01%	83.73%	80.70%
	Geographic Proximity	92.20%	92.56%	90.90%	93.27%	77.07%	78.98%
	Migration	93.44%	94.95%	92.86%	86.03%	73.03%	71.19%
	Trade	91.02%	91.22%	92.03%	93.35%	76.48%	79.32%
Situation Verifying	Democracy	94.88%	91.89%	94.55%	94.11%	85.40%	80.81%
	Geographic Proximity	94.31%	92.18%	93.44%	88.71%	80.92%	80.47%
	Migration	95.17%	90.73%	96.04%	78.49%	80.96%	70.62%
	Trade	93.13%	92.35%	93.91%	91.44%	81.63%	79.25%
Action-Supportive	Democracy	93.28%	92.63%	92.89%	90.26%	89.49%	72.50%
	Geographic Proximity	93.98%	91.92%	94.01%	90.02%	78.83%	71.12%
	Migration	94.50%	92.12%	96.04%	84.68%	78.03%	74.13%
	Trade	93.36%	93.38%	93.91%	90.05%	81.63%	71.02%

Fig. 8.3 Model accuracy tests

8.2.4 Model Validation

The processed Twitter data exhibits some interesting characteristics associated with the social movement. The final sample included a total of 11,876 tweets. After removing the uncategorizable tweets, 4628 of the tweets were ad-hoc reporting, 3572 situation-verifying, and 3676 action-supportive messages There are noticeable differences for each communication type at different periods while overall the three types of tweets show consistent patterns.

First, around the first week of February during which several key events occurred, the volume of action-supportive messages sharply increased, including the brutal violence in Tahrir Square against protesters on February 2 and Cairo's biggest protest on February 4, the so-called "day of departure". Second, on January 27 right after the Egyptian Internet was unblocked, the volume of ad-hoc reporting messages slightly increased. This was followed by a sharp increase around the period of Mubarak's resignation on February 11. Meanwhile, the volume of situation-verifying messages grew steadily throughout the entire period.

Based on three message types, four spatial arrangements, and three time frames, we validated the spatiotemporal model with 36 sub-datasets ($3 \times 4 \times 3 = 36$). Two series of modeling were performed, one with the raw tweet volumes and another with the residual values (computed after accounting for the Internet penetration effect) as a dependent variable respectively. The simulations were performed with a Matlab best fitting program, fminsearch, to obtain the possible accuracy. As seen in Fig. 8.3, the accuracy of the model fit with the raw tweet volumes is above 90%. The accuracy decreases, with 70.62% being the lowest, when the residuals are taken into consideration. The accuracy of the model fit was measured by the difference between predicted value against the actual observed value. While the decline in accuracy suggests that the Internet penetration effect indeed accounted for a good portion of diffusion pattern, the model accuracy seemed generally acceptable.

8.2.5 Discussion

In this study, we aim to extend a spatiotemporal diffusion model to a new partial differential equation model with advection term to understand a protest information diffusion process across networked global public. The new model has the premise that global mobilization of social media publics is not disparate from existing international-relational dynamics. As a result, we proposed to arrange spatial "distances" between the origin protest country (Egypt) and other global regions in various ways according to their physical, cultural, economic, and ideological proximity with the origin region. The analytic focus of this study was on the spatial diffusion process, represented by the parameter associated with the advection term: we defined it as "spatial spread-ability" considering that this parameter represented the rate of change in tweet volumes (or residual volumes) from the left to the right. In the simulations we assume that the local diffusion $\frac{\partial(de^{-bx}I_x)}{\partial x}$ is the same or close for different types of spatial arrangements, and we can compare the value of $g(x)$ to determine how fast information moves along a direction. Our best fitting program reveals that the local diffusions for the four spatial arrangements are the same or close.

The current study presents an innovative approach exploring spatiotemporal diffusion of protest ideas. In particular, it initiates a starting point to use spatio-temporal modeling approach with abstract distances. Given the paucity of mathematical diffusion modeling in the scholarship of social movements and protests, the current study contributes to advance formal explorations of information diffusion patterns. As statistical or mathematical models generally do, a formal modeling allows researchers to predict the future. A model is proposed based on theories and theorems and is fitted to an existing empirical data retrospectively to validate the model accuracy. Then, the parameters computed from the empirical modeling can be applied to predict the trajectory of similar types of events in the future, for which the data may not yet be available. To this end, we need to further category three types of messages because each of them may be associated with different ranges of parameters. Especially, the PDE-based model is a novel approach to information diffusion, simultaneously taking advantage of space and time dimension considerations. Social media have become effective channels for protest mobilization. Leveraging such social media data to explore the ways in which protest ideas and information are diffused through online global networks may contribute to increased understanding of the dynamics underlying spontaneous emergence of collective actions in contemporary socio-digital environment. In summary, the partial differential equation models are promising in studying spatio-temporal social problems.

8.3 A PDE-Based Influenza Surveillance System

We use PDE model in the prototype of a real-time influenza surveillance system to understand and predict the influenza trend by harnessing influenza signals embedded in social media. The Center for Disease Control and Prevention (CDC) monitors influenza-like illness (ILI) cases by collecting data from clinic visits, and it publishes state ILI case counts and regional ILI case counts on the CDC website weekly. There is usually 1 week delay in these ILI reports. Early work on web-based influenza surveillance usually relies on query data from major search engines and focuses on status updates from the microblogging website [22]. Google's Flu Trends makes use of the search queries at Google as well as ILI data to predict flu outbreak. More recently, Twitter has been shown as a reliable source for studying flu trends [1, 2]. Twitter is real-time; therefore it avoids 1-week delay introduced by CDC.

Furthermore, Twitter can provide fine grained location information such as city, famous place, Zip Code, even location coordinates of. Based on the spatio-temporal modeling approach, we have built a new influenza surveillance system applying partial differential models to social media data and aim to effectively facilitate an early detection of influenza outbreaks. Our approach can achieve faster, near real time and localized prediction of the emergence and spread of influenza epidemic.

We will emphasize the integration of the clustering, embedding algorithms and PDE models for improving influenza prediction. We choose two different clustering methods: the higher-order motif clustering discussed in Sect. 4.4 and the bipartite graph clustering described in Sect. 4.5. The purpose of clustering is to find hidden communities (such as frequent air travels) related to flu epidemics. This, in turn, will cause the PDE model to describe flu information diffusion in Twitter more accurately. The system architecture includes components for Twitter data collection, geo-location information extraction, clustering, PDE model development and simulations, validation of the model with previous CDC data, prediction of influenza trends as shown in Fig. 8.4. Some preliminary results are reported in [131].

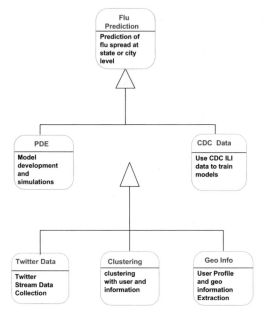

Fig. 8.4 Architecture of a PDE-based influenza surveillance system

8.3.1 Twitter Data

It was shown that there exists significant correlation between Twitter flu tweets and the ILI cases reported by CDC [1, 2]. To further verify the relevance of flu trends modeled by our system, we correlate geo-tagged flu tweets with the reported flu cases released from CDC official statistics. We select the flu tweets collected between January 3 and March 26, 2014, which align with the flu season. Figure 8.5 shows the number of weekly new flu tweets in Twitter and the number of weekly reported ILI cases provided by CDC and it shows strong linear correlation between the lines. To measure the linear correlation, we apply Pearson's product-moment correlation coefficient

$$\rho_{X,Y} = \frac{\text{cov}(X,Y)}{\sigma(X)\sigma(Y)}$$

where $\text{cov}(X,Y)$ is the covariance between variables X and Y, and $\sigma(X), \sigma(Y)$ are the standard deviation of X, Y respectively. The result shows that the correlation coefficient between Twitter weekly flu tweets count and the reported ILI at national level is as high as 0.9297.

Our system collects tweets containing influenza related keywords and the profiles of users involved in these influenza tweets, and users involving in these tweets. Specifically, tweets include information such as username of the user who tweeted this tweet, status, and timestamp. User profile includes information such as following relationship, number of friends, the user profile creation date, location (including Zip Code) and status update.

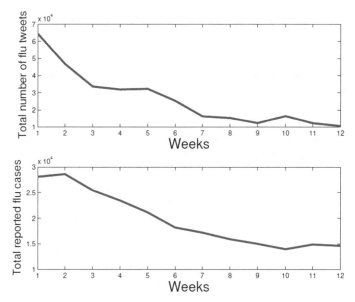

Fig. 8.5 Comparison of the volume of Twitter flu data and number of CDC ILI cases

To extract geo-location information from tweet, we focus on the following four fields: coordinates, place, profile location, and text. It is estimated a large portion of Twitter users include specific location in their user profiles. However, there are still challenges as follows: only a very small percentage of Twitter users add GPS information to their tweets; a significant number of users attempt to obscure their location in their profiles or use valid locations with non-standard spelling. In addition to these mechanisms supported directly by Twitter APIs, we can infer the geographic location associated with Twitter messages along with their followers' retweets. For example, users can provide location in tweets indirectly (for example, "My plane just landed at LAX"). Takhteyeva et al. [118] finds that about 85% of such tweets referred to a real place, at the level of a country or smaller, while 65% named a geographic unit the size of a major metropolitan area or smaller. Our system includes a database containing names of states, abbreviations, cities, and common misspellings for all U.S. states. We have expanded the database to include more fine-grained location information such as Zip Codes, airport codes, monuments, etc. The geo-location information of each tweet is stored as structured objects containing multiple attributes, such as country, city, Zip

Code, neighborhood, and latitude and longitude. This provides finer location granularity than CDC data, which only provides state and regional data.

8.3.2 PDE Modeling Based on Higher-Order Graph Clustering

In this section, we demonstrate an innovative integration of clustering algorithms and PDE modeling. First we use the motif clustering algorithm in Sect. 4.4 to divide the U.S. states into a number of clusters and then use PDE to predict a flu tweet level. To make modeling and reporting more feasible; and to make it easier for the public to understand conditions within the U.S., CDC reports flu numbers at regional levels defined by the U.S. Department of Health and Human Services (HHS), which divides the fifty-two states in United States into ten regions as illustrated in Fig. 8.6a.

To refine the ten regions to further reflect the changes of commuting flows between states, we apply the motif clustering algorithm in Sect. 4.4 to the U.S. state network. We use the 2006–2010 US Census county-to-county commuting flows data to build two separate networks. This data set describes the commuting patterns of individuals between residence and workplace. We first build a county level network that describes commuting flows for individuals between counties. Each node of the country-level network is a (state, county) tuple and its edges are assigned weights corresponding to the flow represented in the data set. From the county-level network it is easy to build a state-level network which provides us with a way to measure how states are connected. Each node of the state level network is a state and its edges are assigned weights based on the aggregated sum of commuting flows across state lines.

We choose, motif M_{13} in Fig. 4.1 in the motif clustering method to find more connected clusters in the state level network. We can observe from Fig. 4.1 that M_{13} indicates commuting flows between states. In this way we identify hidden communities of commuters related to flu epidemics. This, in turn, will cause the PDE model to more accurately describe the diffusion of flu information in Twitter.

When the regions are embedded into a line, they are placed in the x-axis at $x = 1, 2, \ldots 10$ according to their ordering in Fig. 8.6b, c. The ordering in the maps (b) and (c) in Fig. 8.6 is produced by the embedding algorithm based on their connectivity as we discuss in Sect. 5.2. The goal of this embedding process is to place more connected regions closer along the x-axis. First we create the Laplacian from the adjacency matrix of the state level network. Then we calculate its Fiedler vector which is the eigenvector corresponding to the second largest eigenvalue. Now we arrange the regions on the x-axis according to the order of the sorted Fiedler vector. This is done for both the HHS and motif regions. For example, for Fig. 8.6c,

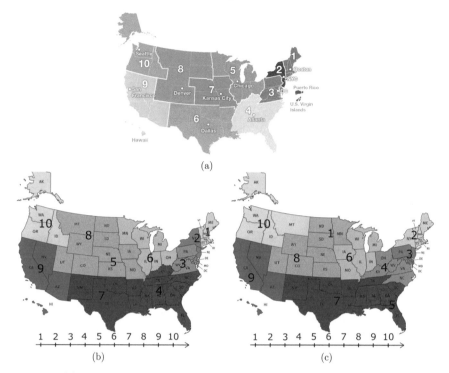

Fig. 8.6 (**a**) HHS ten regions. (**b**) HHS regions with order for embedding. (**c**) Ten regions created by the motif

the sorted Fiedler vector of the ten regions created by the motif clustering is $(-0.18527, -0.14938, -0.14605, -0.13713, -0.13054, -0.11428, -0.10266,$ $-0.06942, 0.115844, 0.918883)$ corresponding to regions $1, 2, 3, \ldots 10$ respectively. The smallest component of the Fiedler vector (-0.18527) corresponds to the least connected region (ND, MD and MN). As a result of the embedding procedure, the order of the regions in Fig. 8.6b is slightly different from that of the original HHS regions in Fig. 8.6a $(7 - >5, 5 - >6, 6 - >7)$.

For simplicity, we choose the following logistic equation as in (5.7.1)

$$\frac{\partial I}{\partial t} = \frac{\partial (de^{-bx} I_x)}{\partial x} + r(t)I(h(x) - \frac{I}{K})$$
$$I(x, 1) = \phi(x), \quad l < x < L \qquad (8.3.1)$$
$$\frac{\partial I}{\partial x}(l, t) = \frac{\partial I}{\partial x}(L, t) = 0, \quad t > 1$$

The PDE simulations were run on the Twitter flu data that were collected for the 2018–2019 flu season (September 30, 2018, though May 18, 2019). We aggregated flu counts with respect to the new regions created by the clustering algorithm and then ordered them by the embedding algorithm.

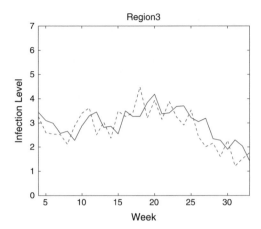

Fig. 8.7 Flu tweet prediction with motif clustering for region 3. Dashed lines are
real data and solid lines are PDE prediction

We normalized the flu data for each region for decreasing the experimental
errors. Then, we use the training data set of 3 weeks to predict the flu level for
the following week, (i.e., we use weeks 1–3, 2–4, 3–5, ... as training data), and
predict the following weeks 4, 5, 6, ... respectively and record the prediction
accuracy for all ten regions at the weeks 4, 5, 6, ... Here we give the detailed
prediction process for week 4 as an example. For the training data set of
weeks 1, 2, 3, we interpolate the discrete data of week 1 in constructing the
initial function $\phi(x)$ of (8.3.1). Next, we use the data of week 2 and week 3 to
train the parameters of the PDE model by the fminsearch function in Matlab.
Finally, we use this PDE model with a new initial function $\phi(x)$ trained by
the data of week 3 to predict the flu level of week 4.

Table 8.1 Prediction accuracy of (8.3.1) with ten HHS and Motif regions

Regions created by	1	2	3	4	5	6	7	8	9	10
HHS	93%	87%	88%	87%	86%	93%	88%	87%	84%	86%
Motif-clustering	91%	91%	92%	92%	89%	92%	90%	90%	88%	91%

Table 8.1 illustrates the average prediction results of regions 1–10 for 30
weeks for the 2018–2019 flu season from September 30, 2018, through May
18, 2019). Clearly PDE with the higher order graph clustering has a better
accuracy for the flu season. It is reasonable to believe that our new region
definition derived from the clustering of the state-level commuter network will
increase the prediction accuracy. Figure 8.7 is a snap shot of PDE simulation
for region 3 with motif clustering method.

8.3.3 PDE Modeling Based on Bipartite Graph Clustering

In this section, we use the co-clustering algorithm in Sect. 4.5 to divide the U.S. states into a number of clusters. We now assume all online users are associated with geo-location information (tags), which could result from user profiles, or tweet contents and other approaches. This association can be conveniently represented by a bipartite graph. Let $\{u_1, ..u_m\}$ be users and $\{t_1, \ldots t_n\}$ be the associated tags, here tags t_i are geo-location information, for example, NY, AZ. Now the co-clustering algorithm can be adopted to cluster the users as well as tags.

To illustrate how this algorithm works, we use the bipartite graph with users $\{u_1, ..u_9\}$ and tags $\{t_1, \ldots t_7\}$ as illustrated in Fig. 8.8 where $t1, t2, t3, \ldots t7$ are related to states, CA, AZ, NM, TX, LA MS, GA respectively. Figure 8.9 is the corresponding adjacency matrix of the network.

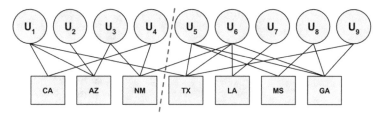

Fig. 8.8 A two-mode network represented as a bipartite graph. The dashed line represents one normalized cut found by spectral clustering

$$A = \begin{array}{c|ccccccc} & t_1 & t_2 & t_3 & t_4 & t_5 & t_6 & t_7 \\ \hline u_1 & 1 & 1 & 0 & 1 & 0 & 0 & 0 \\ u_2 & 0 & 1 & 0 & 0 & 0 & 0 & 0 \\ u_3 & 0 & 1 & 1 & 0 & 0 & 0 & 0 \\ u_4 & 1 & 0 & 1 & 0 & 0 & 0 & 0 \\ u_5 & 0 & 0 & 0 & 1 & 0 & 1 & 1 \\ u_6 & 0 & 0 & 1 & 1 & 1 & 0 & 1 \\ u_7 & 0 & 0 & 0 & 0 & 0 & 1 & 0 \\ u_8 & 0 & 0 & 0 & 0 & 0 & 1 & 1 \\ u_9 & 0 & 0 & 0 & 1 & 0 & 0 & 1 \end{array}$$

Fig. 8.9 Adjacency matrix of the bipartite graph for Fig. 8.8

To use the co-clustering algorithm in Sect. 4.5, we first note that the degrees of u_i and t_j, $D_u = \text{diag}(u_i) = \text{diag}(3, 1, 2, 2, 3, 4, 1, 2, 2)$, $D_t = \text{diag}(t_j) = \text{diag}(2, 3, 3, 4, 1, 3, 4)$. Now we can normalize the adjacency matrix as $\tilde{A} = D_u^{-1/2} A D_t^{-1/2}$ in (8.3.2).

$$\tilde{A} = \begin{pmatrix}
0.4082 & 0.3333 & 0 & 0.287 & 0 & 0 & 0 \\
0 & 0.5774 & 0 & 0 & 0 & 0 & 0 \\
0 & 0.4082 & 0.4082 & 0 & 0 & 0 & 0 \\
0.50 & 0 & 0.4082 & 0 & 0 & 0 & 0 \\
0 & 0 & 0 & 0.2887 & 0 & 0.3333 & 0.2887 \\
0 & 0 & 0.2887 & 0.25 & 0.50 & 0 & 0.25 \\
0 & 0 & 0 & 0 & 0 & 0.5774 & 0 \\
0 & 0 & 0 & 0 & 0 & 0.4082 & 0.3536 \\
0 & 0 & 0 & 0.3536 & 0 & 0 & 0.3536
\end{pmatrix} \qquad (8.3.2)$$

Now compute the singular value decomposition, SVD of $\tilde{A}$, say $\tilde{A} = U\Sigma V^T$. Let $S^{(u)} = U_{2:l}, S^{(t)} = V_{2:l}$, where $l = [\log_2^k] + 1$ if we aim to find k communities. In this case we let $k = 2$. After we obtain $S^{(u)}$ and $S^{(t)}$, the standard clustering algorithm, k-means, can be used to obtain communities. We obtain a joint community indicator of both the user mode and the tag mode as

$$Z = \begin{pmatrix} D_u^{-1/2} S^{(u)} \\ D_t^{-1/2} S^{(t)} \end{pmatrix}$$

For $\tilde{A}$ in (8.3.2), we compute $S^{(u)}$, $S^{(t)}$ and Z as follows

$$S^{(u)} = \begin{pmatrix}
0.3264 \\
0.3467 \\
0.3974 \\
0.3481 \\
-0.3740 \\
-0.0406 \\
-0.3772 \\
-0.4168 \\
-0.1914
\end{pmatrix}, S^{(t)} = \begin{pmatrix}
0.3469 \\
0.5320 \\
0.3303 \\
-0.1034 \\
-0.0229 \\
-0.5787 \\
-0.3761
\end{pmatrix}, Z = \begin{pmatrix}
u_1 & 0.1884 \\
u_2 & 0.3467 \\
u_3 & 0.2810 \\
u_4 & 0.2461 \\
u_5 & -0.2159 \\
u_6 & -0.0203 \\
u_7 & -0.3772 \\
u_8 & -0.2947 \\
u_9 & -0.1353 \\
t_1 & 0.2453 \\
t_2 & 0.3071 \\
t_3 & 0.1907 \\
t_4 & -0.0517 \\
t_5 & -0.0229 \\
t_6 & -0.3341 \\
t_7 & -0.1880
\end{pmatrix}$$

Now it is clear that we can obtain two communities $U_1 = \{u_1, u_2, u_3, u_4\}$ with CA, AZ, NM and $U_2 = \{u_5, u_6, u_7, u_8, u_9\}$ with TX, LA MS, GA, which is exactly the cut as indicated by the dashed line in Fig. 8.8. The first group represents a group at the Western region and the second at the Eastern region.

In our experiment, we select flu tweets from February 13, 2014 to March 26, 2014 with 700 user involved and 8061 tweets in total. The output of this algorithm produces five clusters at $x = 1, 2, 3, 4, 5$. Cluster 2 includes users associated with Connecticut, Mississippi, North Dakota, South Carolina, Utah, and Washington; Cluster 3 is associated with District of Columbia, Illinois, Louisiana, Minnesota, Puerto Rico; Cluster 4 is associated with Florida, Kentucky, Maryland, Wisconsin; Cluster 5 is associated with Idaho, Montana, Rhode Island. Cluster 1, the largest cluster, consists of the rest of states including California, Texas, New York and other states. The clustering result indicates that each cluster has more common interests related to the geolocations respectively. One of possible explanations for the clustering results is that the selected users may travel frequently within each cluster.

This simulation indicates that the PDE model can accurately predict flu tweet counts. The overall accuracy of the simulation for five regions is 89%. In application of this model, we assess epidemic intensity on a real-time basis by examining whether the average of tweet diffusion in Twitter is above a baseline. If simulations of the predictive PDE model indicate that a region experiences an early and sharp increase of relevant tweets, it may be possible to focus resources on that region to identify an outbreak and provide extra vaccine capacity or raise local media awareness as necessary.

8.3.4 Discussion

The prototype system for influenza surveillance aims to make it possible for government or health organizations to raise an alarm for specific regions at those moments when there is a high probability that the flu is breaking out. The focus on local levels in our models will enable a quick local response in preventing the disease from spreading. Our system can identify regions with an early and sharp increase of relevant tweets. The region is highly related to the highest value of $h(x)$ as it represents the influenced rate of information for the group users whose distance from the origin is x. By using the advanced mathematical models, we can calculate the localized posterior likelihood for how related tweets to transit within Twitter based on previous observed data. We use this trend to predict how influenza moves to (or stays at) state, or even city level and in specific population groups.

The system has several limitations. One of the key challenges is to cluster users with geo-location information. Our clustering approach is capable of revealing hidden communities (such as frequent air travelers) related to influenza among online users. Because the ground truth is unknown, it is difficult to verify the clustering results. Nevertheless, we can use previous flu trends and other data to improve our clustering algorithms. The proposed PDE model does not distinguish differences in demographic groups such as age, which is, in particular, significant for influenza outbreaks. As a result, the model parameters and prediction performance will vary for different sections of the population. The accuracy of influenza prediction will depend on the amount of available geocoded data. For tweets with ambiguous locations mentioned in the content, we could consider the recent tweet history from the user who posted the tweet to determine the accurate location. We could enable dynamic location additions to our database based on a location API and structured information from Twitter. Enhanced data mining algorithms could be developed to further augment our location database and alias lists in comparing with other public resources such as Google maps and Wikipedia.

References

1. Achrekar, H., Gandhe, A., Lazarus, R., Yu, S.-H., Liu, B.: Predicting flu trends using twitter data. In: 2011 IEEE Conference on Computer Communications Workshops (INFOCOM WKSHPS), pp. 702–707. IEEE, Piscataway (2011)
2. Achrekar, H., Gandhe, A., Lazarus, R., Yu, S.-H., Liu, B.: Twitter improves seasonal influenza prediction. In: International Conference on Health Informatics (HEALTHINF), pp. 61–70 (2012)
3. Afrouzi, G.A., Brown, K.J.: On principal eigenvalues for boundary value problems with indefinite weight and Robin boundary conditions. Proc. Am. Math. Soc. **127**, 125–130 (1999)
4. Allegretto, W., Huang, Y.X.: A Picone's identity for the p-Laplacian and applications. Nonlinear Anal. **32**, 819–830 (1998)
5. Baptista, R.: Geographical clusters and innovation diffusion. Technol. Forecast. Soc. Chang. **66**, 31–46 (2001)
6. Barber, M.J.: Modularity and community detection in bipartite networks. Phys. Rev. E **76**, 066102 (2007)
7. Barbieri, N., Bonchi, F., Manco, G.: Cascade-based community detection. In: Proceedings of the 6th ACM International Conference on Web Search and Data Mining, pp. 33–42. ACM, New York (2013)
8. Barrat, A., Barthelemy, M., Vespignani, A.: Dynamical Processes on Complex Networks. Cambridge University Press, New York (2008)
9. Benevenuto, F., Rodrigues, T., Cha, M., Almeida, V.: Characterizing user behavior in online social networks. In: Proceedings of the 9th ACM SIGCOMM Conference on Internet Measurement Conference, pp. 49–62. ACM, New York (2009)
10. Bennett, W., Segerberg, A.: The logic of connective action. Inf. Commun. Soc. **15**, 739–768 (2012)
11. Bennett, W., Breunig, C., Givens, T.: Communication and political mobilization: digital media and the organization of anti-Iraq war demonstrations in the US. Polit. Commun. **25**, 269–289 (2008)

© Springer Nature Switzerland AG 2020

H. Wang et al., *Modeling Information Diffusion in Online Social Networks with Partial Differential Equations*, Surveys and Tutorials in the Applied Mathematical Sciences 7, https://doi.org/10.1007/978-3-030-38852-2

12. Benson, A., Gleich, D., Leskovec, J.: Higher-order organization of complex networks. Science **353**, 163–166 (2016)
13. Belkin, M., Niyogi, P.: Laplacian eigenmaps for dimensionality reduction and data representation. Neural Comput. **15**, 1373–1396 (2003). https://doi.org/10.1162/089976603321780317
14. Brauer, F., Castillo-Chvez, C.: Mathematical Models in Population Biology and Epidemiology, 2nd edn. Springer, New York (2012)
15. Brauer, F., Van den Driessche, P., Wu, J.: Mathematical Epidemiology. Springer, Heidelberg (2008)
16. Bruns, A., Highfield, T., Burgess, J.: The Arab Spring and social media audiences English and Arabic Twitter users and their networks. Am. Behav. Sci. **57**, 871–898 (2013)
17. Cantrell, R.S., Cosner, C.: Spatial Ecology via Reaction-diffusion Equations. Wiley, Hoboken (2004)
18. Cha, M., Mislove, A., Adams, B., Gummadi, K.: Characterizing social cascades in flickr. In: Proceedings of the First Workshop on Online Social Networks, pp. 13–18. ACM, New York (2008)
19. Cha, M., Mislove, A., Gummadi, K.: A measurement-driven analysis of information propagation in the flickr social network. In: Proceedings of the 18th International Conference on World Wide Web, pp. 721–730. ACM, New York (2009)
20. Chen, X.F., Friedman, A.: A free boundary problem arising in a model of wound healing. SIAM J. Math. Anal. **32**, 778–800 (2000)
21. Chung, F.: Spectral Graph Theory, vol. 92. American Mathematical Society, Providence (1997)
22. Cook, S., Conrad, C., Fowlkes, A.L., Mohebbi, M.H.: Assessing Google flu trends performance in the United States during the 2009 influenza virus a (H1N1) pandemic. PLoS One **6**, e23610 (2011)
23. Dai, G., Ma, R., Wang, H.: Partial differential equations with robin boundary conditions in online social networks. Discrete Contin. Dyn. Syst. B **20**, 1609–1624 (2015)
24. Dhillon, S.: Co-clustering documents and words using bipartite spectral graph partitioning. In: Proceedings of the 7th ACM SIGKDD International Conference on Knowledge Discovery and Data Mining, pp. 269–274. ACM, New York (2001)
25. Dietz, L.: Inferring shared interests from social networks. In: Proceedings of Neural Information Processing Systems Workshop on Computational Social Science and the Wisdom of Crowds (2010)
26. Dredze, M., Paul, M., Bergsma, S., Tran, H.: Carmen: a Twitter geolocation system with applications to public health. In: AAAI Workshop on Expanding the Boundaries of Health Informatics Using AI (HIAI), pp. 20–24 (2013)
27. Du, Y.H., Lin, Z.G.: Spreading-vanishing dichotomy in the diffusive logistic model with a free boundary. SIAM J. Math. Anal. **42**, 377–405 (2010)

28. Du, Y.H., Ma, L.: Logistic type equations on RN by a squeezing method involving boundary blow-up solutions. J. Lond. Math. Soc. **64**, 107–124 (2001)

29. Easley, D., Kleinberg, J.: Networks, Crowds, and Markets: Reasoning about a Highly Connected World. Cambridge University Press, New York (2010)

30. Evans, L.C.: Partial Differential Equations. Graduate Studies in Mathematics, vol. 19. American Mathematical Society, Providence (1998)

31. Fisher, R.: The wave of advance of advantageous genes. Ann. Eugen. **7**, 355–369 (1937)

32. Fortunato, S.: Community detection in graphs. Phys. Rep. **486**, 75–174 (2010)

33. Gerald, C.F., Wheatley, P.O.: Applied Numerical Analysis. Addison-Wesley, Boston (1994)

34. Ghosh, R., Lerman, K.: A framework for quantitative analysis of cascades on networks. In: Proceedings of the 4th ACM International Conference on Web Search and Data Mining, pp. 665–674. ACM, New York (2011)

35. Girvan, M., Newman, M.: Community structure in social and biological networks. Proc. Natl. Acad. Sci. **99**, 7821–7826 (2002)

36. Granovetter, M.: Threshold models of collective behavior. Am. J. Sociol. **83**, 1420–1443 (1978)

37. Guillaume, J.-L., Latapy, M.: Bipartite graphs as models of complex networks. Phys. A Stat. Theor. Phys. **371**, 795–813 (2006)

38. Guille, A., Hacid, H., Favre, C., Zighed, D.: Information diffusion in online social networks: a survey. SIGMOD Rec. **42**, 17–28 (2013)

39. Guimera, R., Sales-Pardo, M., Amaral, L.: Module identification in bipartite and directed networks. Phys. Rev. E **76**, 036102 (2007)

40. Hagerstrand, T.: Innovation Diffusion as a Spatial Process. University of Chicago Press, Chicago (1968)

41. Hajibagheri, A., Alvari, H., Hamzeh, A., Hashemi, S.: Community detection in social networks using information diffusion. In: Proceedings of the 2012 International Conference on Advances in Social Networks Analysis and Mining (ASONAM 2012), pp. 702–703. IEEE Computer Society, Washington (2012)

42. Heesterbeek, J., Metz, J.: The saturating contact rate in marriage and epidemic models. J. Math. Biol. **31**, 529–539 (1993)

43. Hernandez-Campos, F., Nobel, A.B., Smith, F.D., Jeffay, K.: Statistical clustering of Internet communication patterns. In: Proceedings of Symposium on the Interface of Computing Science and Statistics (2003)

44. Hess, V.: Periodic Parabolic Boundary Value Problems and Positivity. Longman Scientific & Technical, Harlow (1991)

45. Horn, R., Johnson C.: Matrix Analysis, 2nd edn. Cambridge University Press, Cambridge (2013)

46. http://en.wikipedia.org/wiki/Twitter

47. Ikeda, Y., Hasegawa, T., Nemoto, K.: Cascade dynamics on clustered network. J. Phys. Conf. Ser. **221**, 012005 (2010)
48. Ince, E.L.: Ordinary Differential Equation. Dover, New York (1927)
49. Jesus, R., Schwartz, M., Lehmann, S.: Bipartite networks of Wikipedia articles and authors: a meso-level approach. In: Proceedings of the 5th International Symposium on Wikis and Open Collaboration, p. 5. ACM, New York (2009)
50. Jiang, J., Wilson, C., Wang, X., Huang, P., Sha, W., Dai, Y., Zhao, B.: Understanding latent interactions in online social networks. ACM Trans. Web **7**, 18 (2013)
51. Jin, F., Dougherty, E., Saraf, P., Cao, Y., Ramakrishnan, N.: Epidemiological modeling of news and rumors on Twitter. In: Proceedings of the 7th Workshop on Social Network Mining and Analysis, p. 8. ACM, New York (2013)
52. Karypis, G.: CLUTO: A clustering toolkit. TR-02–017, Department of Computer Science, University of Minnesota (2002). http://www.cs.umn. edu/cluto
53. Katz, E.: The two-step flow of communication: an up-to-date report on a hypothesis. Public Opin. Q. **21**, 61–78 (1957)
54. Katz, E., Lazarsfeld, P.F.: Personal Influence: The Part Played by People in the Flow of Mass Communications. Transaction Publishers, Piscataway (2006)
55. Kawamoto, T.: A stochastic model of tweet diffusion on the Twitter network. Phys. A **392**, 3470–3475(2013)
56. Kempe, D., Kleinberg, J.M., Tardos, E.: Maximizing the spread of influence through a social network. In: Proceedings of the 9th SIGKDD International Conference on Knowledge Discovery and Data Mining, pp. 137–146. ACM, New York (2003)
57. Kermack, W.O., McKendrick, A.G.: A contribution to the mathematical theory of epidemics. Proc. R. Soc. Lond. B **115**, 700–721 (1927)
58. Kolmogorov, A., Petrovsky, N.I.: Piscounov, Etude de lequation de la diffusion aveccroissance de la quantite de matiere et son application a un probleme biologique. Bull. Moscow Univ. Math. Mech. **1**, 1–26 (1937)
59. Kreith, K.: Picone's identity and generalizations. Rend. Mat. **8**, 251–261 (1975)
60. Krishnamurthy, B., Wang, J.: On network-aware clustering of web clients. In: Proceedings of the Conference on Applications, Technologies, Architectures, and Protocols for Computer Communication, pp. 97–110. ACM, New York (2000)
61. Kumar, R., Novak, J., Tomkins, A.: Structure and evolution of online social networks. In: Proceedings of the 12th ACM SIGKDD International Conference on Knowledge Discovery and Data Mining, pp. 20–23. ACM, New York (2006)

62. Kwak, H., Choi, Y., Eom, Y.-H., Jeong, H., Moon, S.: Mining communiteis in networks: a solution for consistency and its evaluation. In: Proceedings of the 9th SIGCOMM Conference on Internet Measurement Conference, pp. 301–314. ACM, New York (2009)

63. Kwon, K., Wang, H., Xu, W., Raymond, R.: A spatiotemporal model of Twitter information diffusion: an example of Egyptian revolution 2011. In: Proceedings of Social Media and Society, ACM International Conference Proceeding Series (ICPS), July 27–29, Toronto (2015)

64. Kwon, K., Xu, W., Wang, H., Chon, J.: Spatiotemporal diffusion modeling of global mobilization in social media: the case of Egypt revolution 2011. Int. J. Commun. **10**, 73–97 (2016)

65. Langa, J., Robinson, J., Rodriguez-Bernal, A., Suarez, A.: Permanence and asymptotically stable complete trajectories for nonautonomous lotka-volterra models with diffusion. SIAM J. Math. Anal. **40**, 2179–2216 (2009)

66. Langa, J.A., Bernal, A.R., Suárez, A.: On the long time behavior of non-autonomous Lotka-Volterra models with diffusion via the sub-supertrajectory method. J. Differ. Equ. **249**, 414–445 (2010)

67. Lei, C., Lin, Z., Wang, H.: The free boundary problem describing information diffusion in online social networks. J. Differ. Equ. **254**, 1326–1341 (2013)

68. Lerman, K., Ghosh, R.: Information contagion: an empirical study of spread of news on Digg and Twitter social networks. In: Proceedings of International Conference on Weblogs and Social Media (ICWSM) (2010)

69. Leskovec, J., Mcglohon, M., Faloutsos, C., Glance, N., Hurst, M.: Cascading behavior in large blog graphs. In: SIAM International Conference on Data Mining (SDM), pp. 551–556 (2007)

70. Li, X., Guo, L., Zhao, Y.: Tag-based social interest discovery. In: Proceedings of the 17th International Conference on World Wide Web, pp. 675–684. ACM, New York (2008)

71. Lin, Z.G.: A free boundary problem for a predator-prey model. Nonlinearity **20**, 1883–1892 (2007)

72. Liu, J., Aggarwal, C., Han, J.: On integrating network and community discovery. In: Proceedings of International Conference on Web Search and Data Mining (WSDM), pp. 117–126. ACM, New York (2015)

73. Livne, A., Simmons, M., Adar, E., Adamic, L.: The party is over here: structure and content in the 2010 election. In: Proceedings of the Fifth International AAAI Conference on Weblogs and Social Media, pp. 17–21 (2011)

74. Logan, J.D.: Transport Modeling in Hydrogeochemical Systems. Spinger, New York (2001)

75. Logan, J.D.: Applied Partial Differential Equations. Springer, Berlin (2015)

76. Lou, Y.: Some challenging mathematical problems in evolution of disperal and population dynamics. In: Tutorials in Mathematical Biosciences IV, pp. 171–205. Springer, Heidelberg (2008)

77. Madzvamuse, A., Gaffney, E.A., Maini, P.K.: Stability analysis of nonautonomous reaction-diffusion. J. Math. Biol. **61**, 133–164 (2010)

78. Mahajan, V., Peterson, R. (Eds.): Models for Innovation Diffusion, vol. 48. Sage, Newbury Park (1985)

79. Malliaros, F., Vazirgiannis, M.: Clustering and community detection in directed networks: a survey. Phys. Rep. **533**, 95–142 (2013)

80. Mena-Lorca, J., Hethcote, H.W.: Dynamic models of infectious diseases as regulators of population sizes. J. Math. Biol. **30**, 693–716 (1992)

81. Mierczyn'ski, J.: The principal spectrum for linear nonautonomous parabolic PDEs of second order: basic properties. J. Differ. Equ. **168**, 453–476 (2000)

82. Murray, J.D.: Mathematical Biology I: An Introduction. Springer, New York (1989)

83. Myers, S., Leskovec, J.: Clash of the contagions: cooperation and competition in information diffusion. In: IEEE 12th International Conference on Data Mining (ICDM'12), pp. 539–548 (2012)

84. Myers, S., Zhu, C., Leskovec, J.: Information diffusion and external influence in networks. In: Proceedings of the 18th ACM SIGKDD International Conference on Knowledge discovery and Data Mining, pp. 33–41. ACM, New York (2012)

85. Nazir, A., Raza, S., Chuah, C.-N.: Unveiling facebook: a measurement study of social network based applications. In: Proceedings of the 8th ACM SIGCOMM Conference on Internet Measurement, pp. 43–56. ACM, New York (2008)

86. Nazir, A., Raza, S., Gupta, D., Chuah, C.-N., Krishnamurthy, B.: Network level footprints of facebook applications. In: Proceedings of the 9th ACM SIGCOMM Conference on Internet Measurement Conference, pp. 63–75. ACM, New York (2009)

87. Neatherlina, J., et al.: Influenza A (H1N1) pdm09 during air travel. Travel Med. Infect. Dis. **11**, 110–118 (2013)

88. Nematzadeh, A., Ferrara, E., Flammini, A., Ahn, Y.: Optimal network modularity for information diffusion. Phys. Rev. Lett. **113**, 088701 (2014)

89. Newman, M.: The structure and function of complex networks. SIAM Rev. **45**, 167–256 (2003)

90. Newman, M.: Networks: An Introdution. Oxford University Press, Oxford (2010)

91. Newman, M., Strogatz, S., Watts, D.: Random graphs with arbitrary degree distributions and their applications. Phys. Rev. E **64**, 026118 (2001)

92. Ng, A., Jordan, M., Weiss, Y.: On spectral clustering: analysis and an algorithm. Adv. Neural Inf. Proces. Syst. **2**, 849–856 (2002)

93. Oh, O., Kwon, K., Rao, H.: An exploration of social media in extreme events: rumor theory and Twitter during the Haiti earthquake 2010. In: Proceedings of the International Conference on Information Systems, St. Louis, pp. 12–15 (2010)

94. Pao, C.V.: Nonlinear Parabolic and Elliptic Equations. Plenum Press, New York (1992)

95. Papadopoulos, S., Kompatsiaris, Y., Vakali, A., Spyridonos, P.: Community detection in social media. Data Min. Knowl. Disc. **24**, 515–554 (2011)

96. Peng, C., Xu, K., Wang, F., Wang, H.: Predicting information diffusion initiated from multiple sources in online social networks. In: 2013 Sixth International Symposium on Computational Intelligence and Design (ISCID), pp. 96–99. IEEE, Piscataway (2013)

97. Porter, M., Onnela, J., Mucha, P.: Communities in networks. Not. Am. Math. Soc. **56**, 1082–1097 (2009)

98. Protter, M., Weinberger, H.: Maximum Principles in Differential Equations. Springer, New York (1984)

99. Radford, S.: Linking innovation to design: consumer responses to visual product newness. J. Prod. Innov. Manag. **28**, 208–220 (2011)

100. Ren, Z., Shao, F., Liu, J., Guo, Q., Wang, B.: Node importance measurement based on the degree and clustering coefficient information. Acta Phys. Sin. **62**, 128901 (2013)

101. Ren, J., Zhu, D., Wang, H.: Spreading-vanishing dichotomy in information diffusion in online social networks with intervention. Discrete Contin. Dynam. Syst. B **24**, 1843–1865 (2019)

102. Rodriguez-Bernal, A., Vidal-López, A.: Existence, uniqueness and attractivity properties of positive complete trajectories for non-autonomous reaction-diffusion problem. Discrete Contin. Dynam. Syst. **18**, 537–567 (2007)

103. Rodriguez-Bernal, A., Vidal-López, A.: Extremal equilibria for reaction-diffusion equations in bounded domains and applications. J. Differ. Equ. **244**, 2983–3030 (2008)

104. Rogers, E.M.: Diffusion of Innovations, 4th edn. Free Press, New York (1995)

105. Romero, D., Meeder, B., Kleinberg, J.: Differences in the Mechanics of Information Diffusion Across Topics: Idioms, Political Hashtags, and Complex Contagion on Twitter. In: Proceedings of 20th International World Wide Web Conference (2011)

106. Romero, D., Tan, C., Ugander, J.: On the interplay between social and topical structure. In: Proceedings of the 7th International AAAI Conference on Weblogs and Social Media (ICWSM) (2013)

107. Ruan, Y., Fuhry, D., Parthasarathy, S.: Efficient community detection in large networks using content and links. In: Proceedings of International Conference on World Wide Web (WWW), pp. 1089–1098 (2013)

108. Rubinstein, L.I.: The Stefan Problem. AMS Translations, vol. 27. American Mathematical Society, Providence (1971)
109. Ryan, B., Gross, N.C.: The diffusion of hybrid seed corn in two Iowa communities. Rural. Soc. **8**, 15–24 (1943)
110. Sawardecker, E., Amundsen, C., Sales-Pardo, M., Amaral, L.: Comparison of methods for the detection of node group membership in bipartite networks. Eur. Phys. J. B Condense Matter Complex Syst. **72**, 671–677 (2009)
111. Schneider, F., Feldmann, A., Krishnamurthy, B., Willinger, W.: Understanding online social network usage from a network perspective. In: Proceedings of the 9th ACM SIGCOMM Conference on Internet Measurement Conference, pp. 35–48. ACM, New York (2009)
112. Schwartz, M., Wood, D.: Discovering shared interests using graph analysis. Commun. ACM **36**, 78–89(1993)
113. Shafaei, M., Jalili, M.: Community structure and information cascade in signed networks. N. Gener. Comput. **32**, 257–269 (2014)
114. Smith, H.: Monotone Dynamical Systems: An Introduction to the Theory of Competitive and Cooperative Systems. American Mathematical Society, Providence (1995)
115. Smith, J., Wiest, D.: The uneven geography of global civil society: national and global influences on transnational association. Soc. Forces **84**, 621–652 (2005)
116. Szomszor, M., Alani, H., Cantador, I., O'Hara, K., Shadbolt, N.: Semantic Modelling of User Interests Based on Cross-Folksonomy Analysis. Springer, Heidelberg (2008)
117. Takahashi, L., Maidana, N., Ferreira, W., Pulino, P., Yang, H.: Mathematical models for the Aedes aegypti dispersal dynamics: travelling waves by wing and wind. Bull. Math. Biol. **67**, 509–528 (2005)
118. Takhteyeva, Y., Gruzdb, A., Wellman, B.: Geography of Twitter networks. Soc. Netw. **34**, 73–81 (2012)
119. Tang, Q., Lin, Z.: The asymptotic analysis of an insect dispersal model on a growing domain. J. Math. Anal. Appl. **378**, 649–656 (2011)
120. Tang, L., Liu, H.: Community Detection and Mining in Social Media. Morgan & Claypool, San Rafael (2010)
121. Tang, S., Blenn, N., Doerr, C., Van Mieghem, P.: Digging in the Digg social news website. IEEE Trans. Multimedia **13**, 1163–1175 (2011)
122. Tatem, A., et al.: Mapping populations at risk: improving spatial demographic data for infectious disease modeling and metric derivation. Popul. Health Metr. **10**, 8 (2012)
123. Tufekci, Z.: Big data: pitfalls, methods and concepts for an emergent field (2013). http://dx.doi.org/10.2139/ssrn.2229952
124. Ver Steeg, G., Ghosh, R., Lerman, K.: What stops social epidemics? In: Proceedings of the 5th International AAAI Conference on Weblogs and Social Media (ICWSM) (2011)

125. Wang, H.: Spreading speeds and traveling waves for non-cooperative reaction-diffusion systems. J. Nonlinear Sci. **21**, 747–783 (2011)

126. Wang, H., Castillo-Chavez, C.: Spreading speeds and traveling waves for non-cooperative integro-difference systems. Discrete Contin. Dyn. Syst. B **17**, 2243–2266 (2012)

127. Wang, X.-S., Wang, H., Wu, J.: Traveling waves of diffusive predator-prey systems: disease outbreak propagation. Discrete Contin. Dyn. Syst. A **32**, 3303–3324 (2012)

128. Wang, F., Wang, H., Xu, K.: Diffusive logistic model towards predicting information diffusion in online social networks. In: 2012 32nd International Conference on Distributed Computing Systems Workshops (ICDCSW), pp. 133–139. IEEE, Piscataway (2012). https://doi.org/10.1109/ICDCSW.2012.16

129. Wang, F., Xu, K., Wang, H.: Discovering shared interests in online social networks. In: 2012 32nd International Conference on Distributed Computing Systems Workshops (ICDCSW). pp. 163–168. IEEE, Piscataway (2012). https://doi.org/10.1109/ICDCSW.2012.15

130. Wang, F., Wang, H., Xu, K., Wu, J., Xia, J.: Characterizing information diffusion in online social networks with linear diffusive model. In: 2013 33nd International Conference on Distributed Computing Systems (ICDCS), pp. 307–316. IEEE, Piscataway (2013). https://doi.org/10.1109/ICDCS.2013.14

131. Wang, F., Wang, H., Xu, K., Raymond, R., Chon, J.,Fuller, S., Debruyn, A.: Regional level influenza study with geo-tagged Twitter data. J. Med. Syst. **40**, 189 (2016)

132. Wang, Y., Wang, H., Zhang, S.: A weighted higher-order network analysis of fine particulate matter (PM2.5) transport in Yangtze river delta. Phys. A **496**, 654–662 (2018)

133. Wayant, N., Crooks, A., Stefanidis, A., Croitoru, A., Radzikowski, J., Stahl, J., Shine, J.: Spatiotemporal clustering of Twitter feeds for activity summarization. In: International Conference on Geographic Information Science, pp. 1–6 (2012)

134. Wei, S., Mirkovic, J., Kissel, E.: Profiling and clustering internet hosts. In: Proceedings of the International Conference on Data Mining, pp. 269–275 (2006)

135. Weinberger, H., Lewis, M., Li, B.: Analysis of linear determinacy for spread in cooperative models. J. Math. Biol. **45**, 183–218 (2002)

136. Weiss, Y.: Segmentation using eigenvectors: a unifying view. In: Proceedings of the 7th International Conference on Computer Vision, pp. 975–982. IEEE, Piscataway (1999)

137. Xu, K., Wang, F., Gu, L.: Network-aware behavior clustering of Internet end hosts. In: 2011 Proceedings IEEE INFOCOM, pp. 2078–2086. IEEE, Piscataway (2011)

138. Xu, K., Wang, F., Jia, X., Wang, H.: The impact of sampling on big data analysis of social media: a case study on Flu and Ebola. In: IEEE Global Communications Conference (GLOBECOM), pp. 1–6 (2015)

139. Yang, J., Counts, S.: Comparing information diffusion structure in weblogs and microblogs. In: Proceedings of the 4th International AAAI Conference on Weblogs and Social Media, pp. 351–354 (2010)

140. Yang, J., Leskovec, J.: Modeling information diffusion in implicit networks. In: 2010 10th International Conference on Data Mining (ICDM), pp. 599–608. IEEE, Piscataway (2010)

141. Yu, B., Fei, H.: Modeling social cascade in the Flickr social network. In: Proceedings of International Conference on Fuzzy Systems and Knowledge Discovery (FSKD), vol. 7, pp. 566–570 (2009)

142. Zafarani, R., Abbasi, M., Liu, H.: Social Media Mining: An Introduction. Cambridge University Press, Cambridge (2014)

143. Zanardi, V., Capra, L.: Social ranking: uncovering relevant content using tag-based recommender systems. In: Proceedings of the 2008 ACM Conference on Recommender Systems, pp. 51–58. ACM, New York (2008)

144. Zhang, L., Zhong, X., Wan, L.: Modeling structure evolution of online social networks. In: 2012 8th International Conference on Information Science and Digital Content Technology (ICIDT), pp. 15–19. IEEE, Piscataway (2012)

145. Zhang, X., Sun, G.-Q., Zhu, Y.-X., Ma, J., Jin, Z.: Epidemic dynamics on semi-directed complex networks. Math. Biosci. **246**, 242–251 (2013)

146. Zhang, Z., Liu, C., Zhan, X., Lu, X., Zhang, C., Zhang, Y.: Dynamics of information diffusion and its applications on complex networks. Phys. Rep. **651**, 1–34 (2016)

147. Zhu, L., Zhao, H., Wang, H.: Bifurcation and control of a delayed diffusive logistic model in online social networks. In: Proceedings of the 33rd Chinese Control Conference, Nanjing (2014)

148. Zhu, L., Zhao, H., Wang, H.: Complex dynamic behavior of a rumor propagation model with spatial-temporal diffusion terms. Inf. Sci. **349–350**, 119–138 (2016)

149. Zhu, L., Zhao, H., Wang, H.: Stability and spatial patterns of an epidemic-like rumor propagation model with diffusions. Phys. Scr. **94**, 085007 (2019)

150. Zhu, L., Zhao, H., Wang, H.: Partial differential equation modeling of rumor propagation in complex networks with higher order of organization. Chaos **29**, 053106 (2019)

Index

Symbols
S-shaped curves, 5
k-clique percolation, 41
k-means algorithm, 39
n-hop neighborhood, 28

A
ad-hoc reporting, 114
adjacency matrix, 29, 129
agglomerative clustering algorithm, 23

B
bifurcation, 82
bipartite graphs, 21, 22
bipartition, 30

C
collective action-supportive, 114
comparison principle, 74
competition systems, 65
cooperative systems, 63, 107
cubic splines interpolation, 52
cut, 30
cyber-distances, 17
cycle, 28

D
degree, 28
diffusion-advection, 117

diffusion of innovations, 3, 4
diffusive logistic model, 50
directed graphs, 28

E
edges, 23, 28
Egyptian revolution, 113
eigenvalue problem, 85
eigenvector, 31
external-influence diffusion model, 9

F
flux, 47
free boundary problems, 70
friendship hops, 17

G
generalized eigenvalue problem, 31, 38
graph partitioning, 41
Gronwall inequality, 86

I
in-degrees, 28
incidence matrix, 29
indefinite weight, 82
influenced users, 19
internal-influence diffusion, 9

K
Kermack-McKendrick SIR model, 13

© Springer Nature Switzerland AG 2020
H. Wang et al., *Modeling Information Diffusion in Online Social Networks with Partial Differential Equations*, Surveys and Tutorials in the Applied Mathematical Sciences 7, https://doi.org/10.1007/978-3-030-38852-2

L
Laplacian matrix, 29
left and right singular vectors, 39
linear diffusive model, 53
logistic adoption curve, 5
logistic curve, 5
logistic equation, 50
Lotka-Volterra competition, 65

M
minimum wave speed, 110
mixed-influence diffusion, 10
mixed-influence diffusion model, 9
multiple-sources, 60

N
Neumann boundary condition, 52
node, 28
normalized cut, 32

O
one-mode projection, 23
one-mode projection graph, 23
one-mode projection graphs, 23
out-degrees, 29

P
partial differential equation, 113
partition vector, 30, 31

R
ratio-cut, 31

S
shared interests, 20
shortest path, 28
similarity matrix, 23
singular value decomposition, 38
situation verifying, 114
social contagion, 41
social graph, 27
social network, 27
spatial spread-ability, 122
spectral clustering methods, 41
spreading speeds, 98
stability, 82
standard incidence, 12

T
the Fiedler vector, 30
traveling wave solutions, 98
two-step flow theory, 6

U
undirected graph, 28
upper and lower solutions, 83
user-by-tag matrix, 38

V
vertex weight matrix, 31
voting consistency, 24

Reigh Quotient, 30
reaction-diffusion equations, 90
right-traveling waves, 119
Robin boundary conditions, 82

Printed in the United States
By Bookmasters